绘画与建筑
Painting and Architecture
Wang Yun
王昀

中国电力出版社
CHINA ELECTRIC POWER PRESS

图书在版编目（CIP）数据

跨界设计：绘画与建筑 / 王昀著.–北京：中国电力出版社，2016.6
ISBN 978-7-5123-9291-5

Ⅰ.①跨... Ⅱ.①王... Ⅲ.①建筑设计 Ⅳ.①TU2

中国版本图书馆CIP数据核字(2016)第092003号

感谢北京建筑大学建筑设计艺术研究中心项目的支持

内容提要

本书通过一系列揭示绘画与建筑之间关系的图像操作，阐明绘画与建筑之间所存在的于空间操作层面上的结合点，希冀能够为艺术与建筑之间重新建立起互动关系，为现代艺术确立新的价值的同时，能为建筑设计提供新的视野和构思途径。本书适合建筑、设计、艺术专业的师生及建筑与艺术爱好者阅读。

中国电力出版社出版发行
北京市东城区北京站西街19号　100005
http://www.cepp.sgcc.com.cn
责任编辑：王　倩
封面设计：方体空间工作室（Atelier Fronti）
版式设计：王风雅
责任印制：蔺义舟
责任校对：李　楠
英文翻译：孙　炼
北京盛通印刷股份有限公司印制·各地新华书店经售
2016年6月第1版·第1次印刷
787mm×1092mm 1/16·13.75 印张·320千字
印数：1–2000册
定价：48.00元

敬告读者
本书封底贴有防伪标签，刮开涂层可查询真伪
本书如有印装质量问题，我社发行部负责退换
版权专有　翻印必究

Abstract

Through a series of image operations revealing relationship between painting and architecture, this book aims to demonstrate space combination existed in the two, in order to reconstruct interaction, establish new value for modern art, and at the meantime, provide new horizon and inspirations for architecture design. This book is for architecture, design, and art researchers, also serves for architecture and art lovers.

序　Preface

　　直到19世纪为止的很长时间，绘画一直是以宗教和故事题材作为主要表现对象，绘画与建筑之间的关系也一直将绘画本身视作一种建筑装饰物，绘制或悬挂在建筑的内外墙面，协助建筑展示某种风景或作为诉诸某种宗教含义的视觉呈现。尽管古典绘画的后期，著名画家库尔贝曾誓言再也不做绘制在建筑天花上的天使绘画了，但他的画风也不过是从以往描写宗教题材转为对现实中农夫等现实题材的描绘，就绘画本身而言，本质上依然是对某种对象物来进行绘制和描写。进入20世纪，立体派的出现，特别是康定斯基发现了绘画中纯粹的色彩、形式组合是构成美的原理而非画面本身的具象形态及意义本身，于是采用几何学形态组成的绘画与描绘具象对象物形态的绘画在艺术领域开始获得了等同的价值。也正因如此，在我看来：与几何学密切相关的建筑能够并应该在这个结点上与绘画本身产生关联。沿着这样的思考，我们便开始将绘画与建筑再次加以关联，通过一系列研究性试做，希冀并试图重新找回绘画与建筑之间的结合点，试图找回绘画本身，确切地讲，是现代绘画本身，与现代建筑之间存在的能够重新建立起关联性的价值，这种关联性的价值，我想将不再会是那些仅仅停留在古典绘画与建筑之间所存在的那种建立在具象的和装饰层面上的价值，我们所希冀唤回的，应该是绘画与建筑之间存在于空间与身体性层面的价值。

Not until 19th century, religious stories had long been the principle object of painting. Speaking of relationship, painting is regarded as architecture decorations - painted or hung on interior or exterior walls of buildings, to visually assist architecture showcasing some view or religious implications. Although the famous painter Courbet vowed to never paint angels on building ceilings in latter phase of classical painting, his style was merely changed from religious to practical themes, such as farmers. Painting per-se, is still some kind of drawing or depiction of certain subjects. Cubism was born in 20th Century, especially since Kandinsky discovered the pure color-form combination is principle of beauty in a painting, instead of the concrete form and meaning. Therefore, geometry painting and concrete subject painting began to share equal value in the art field. For the exact same reason, architecture, which is closely related to geometry, should be able to connect with painting at this point. Thus, we started to relate painting to architecture once again. Through a series of research experiments expecting to recover combination between Painting and Architecture, and regain painting's own value. And to be precise, it's the value of which modern painting itself can rebuild relevance with modern architecture. I think, the value of relevance will not only lie in the concrete and decoration level of classical painting and architecture, but more to expect, in the spatial and corporal level.

王昀
Wang Yun
2016年02月

目录

序

导读

0 绘画与建筑关系的思考   1

1 绘画中的空间与建筑中的空间   9
  实例 1 《码头和海》绘画的空间性表达   12
  实例 2 《开花的苹果树》绘画的空间性表达   16
  实例 3 《有线条的构图（黑和白的构图）》绘画的空间性表达   20
  实例 4 《有色块的构图 B》绘画的空间性表达   24
  实例 5 《色彩平衡构成三号》绘画的空间性表达   28
  实例 6 《构成 4 号（离开工厂）》绘画的空间性表达   34
  实例 7 《纽约市 I》绘画的空间性表达   40
  实例 8 《俄罗斯舞蹈的韵律》绘画的空间性表达   48
  实例 9 《圆形画 51》绘画的空间性表达   54
  实例 10 《爱国庆祝会（自由单词绘画）》绘画的空间性表达   58
  实例 11 《男人的头》绘画的空间性表达   64
  实例 12 《抚摸黑人妇女胸部的一颗星》绘画的空间性表达   72
  实例 13 《蓝色二号》绘画的空间性表达   76
  实例 14 《被蓝色光环围绕着的云雀的一只翅膀，正伸向睡在草地上的罂粟的心脏》绘画的空间性表达   82
  实例 15 《爵士乐：「礁湖」的插图》绘画的空间性表达   86
  实例 16 《爱斯基摩人》绘画的空间性表达   92
  实例 17 《头发飘扬的裸像》绘画的空间性表达   100

实例 18 《哈马马特的主题》绘画的空间性表达 108
实例 19 《富饶国土上的纪念碑》绘画的空间性表达 114
实例 20 《傍晚的火》绘画的空间性表达 120
实例 21 《岩石间的小镇》绘画的空间性表达 124
实例 22 《蓝色的夜》绘画的空间性表达 130
实例 23 《和谐的战场》绘画的空间性表达 134
实例 24 《构成第八号》绘画的空间性表达 138
实例 25 《黄·红·蓝》绘画的空间性表达 144
实例 26 《粉红色的音调》绘画的空间性表达 150
实例 27 《十三个矩形》绘画的空间性表达 154
实例 28 《难以忍受的张力》绘画的空间性表达 158
实例 29 《无缘无故的上升》绘画的空间性表达 162
实例 30 《画了节拍器的静物》绘画的空间性表达 168
实例 31 《小提琴》绘画的空间性表达 174
实例 32 《有玻璃杯和报纸的静物》绘画的空间性表达 180

2 空间的叠加的示例 187
实例 33 《狐步舞 A》《有黑色线条的构图 II》《有两条线的构图》绘画叠加的空间性表达 188

3 思考的延伸 197

Contents

Preface

Introduction

0 Thoughts on correlation between painting and architecture — 1

1 Spaces in painting and architecture — 9
    Sample 1 *Pier and Ocean*'s spatial expression — 12
    Sample 2 *The Flowering Apple Tree*'s spatial expression — 16
    Sample 3 *Composition with Lines (Composition in Black and White)*'s spatial expression — 20
    Sample 4 *Composition in Color B*'s spatial expression — 24
    Sample 5 *Composition III with Colored Planes*' spatial expression — 28
    Sample 6 *Composition No.4 (Leaving the Factory)*'s spatial expression — 34
    Sample 7 *New York City I*'s spatial expression — 40
    Sample 8 *Rhythms of a Russian Dance*'s spatial expression — 48
    Sample 9 *Tondo 51*'s spatial expression — 54
    Sample 10 *Interventionist Demonstration (Free Word Painting)*'s spatial expression — 58
    Sample 11 *Man's Head*'s spatial expression — 64
    Sample 12 *A Star Caresses the Breast of a Black Woman*'s spatial expression — 72
    Sample 13 *Blue II*'s spatial expression — 76
    Sample 14 *The Wing of the Lark, Aureoled by the Blue of Gold, Reaches the Heart of the Poppy Sleeping on the Grass Adorned with Diamonds*' spatial expression — 82
    Sample 15 *Illustration for Jazz: Le Lagon*'s spatial expression — 86
    Sample 16 *The Eskimo*'s spatial expression — 92
    Sample 17 *Nude with Flowing Hair*'s spatial expression — 100

Sample 18	*Motif from Hamamet*'s spatial expression	108
Sample 19	*Monument in Fertile Country*'s spatial expression	114
Sample 20	*Fire in Evening*'s spatial expression	120
Sample 21	*Small Town among the Rocks*' spatial expression	124
Sample 22	*Blue Night*'s spatial expression	130
Sample 23	*Harmonized Battle*'s spatial expression	134
Sample 24	*Composition VIII*'s spatial expression	138
Sample 25	*Yellow-Red-Blue*'s spatial expression	144
Sample 26	*Accent in Pink*'s spatial expression	150
Sample 27	*Thirteen Rectangles*' spatial expression	154
Sample 28	*Hard Tension*'s spatial expression	158
Sample 29	*Gratuitous Ascent*'s spatial expression	162
Sample 30	*Still Life with Metronome*'s spatial expression	168
Sample 31	*The Violin*'s spatial expression	174
Sample 32	*Still Life with Glass and Newspaper*'s spatial expression	180

2 Spatial superposition sample   187

    Sample 33 Juxtaposition spatial expression of   188
                    Fox Trot A, *Composition II with Black Lines*, and *Lozenge Composition with Two Lines*

3 Further thoughts   197

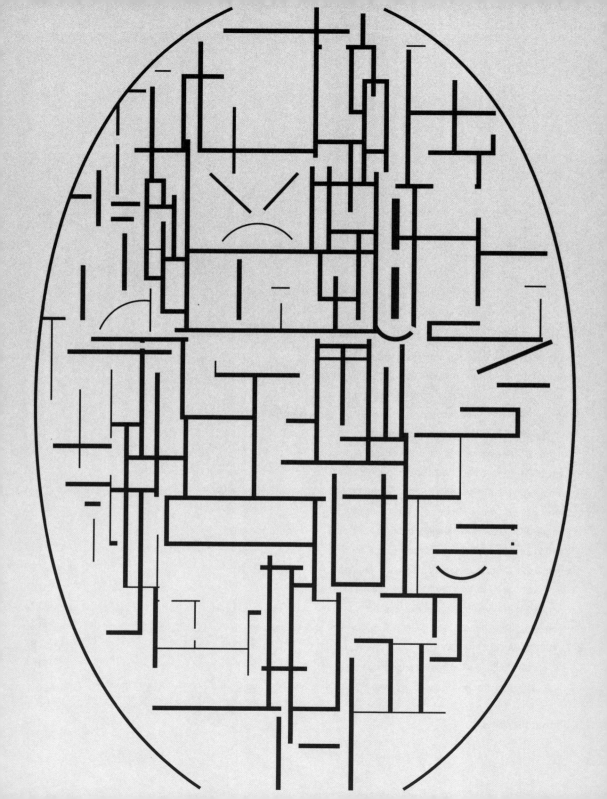

绘画与建筑都使用点、线、面来进行表达
Painting and architecture both use dots, lines, and planes as expression approaches

左图：从《有淡色块的圆形构图》绘画中抽取的空间组成图
Left: Spatial composition diagram extracted from *Oval Composition with Light Colors*

导读 Introduction

在绘画中，点、线、面的关系，或许只是作为一种图案的构成方式来进行操作，但是，在建筑学的范畴里，点、线、面却是一系列构成空间的示意符号。当建筑师面对表面看上去是由点、线、面构成的"图案"的图纸时，其本人从中所读取的一定是一系列拥有空间和功能含义的内容，也是包含有建造指向的内容。譬如："点"可能指向柱子，"线"在图纸中可能指向墙，而"面"可能表示屋顶……于是，当建筑师面对点、线、面所构成的"图案"时，其自身瞬间产生的是三维层面的理解。确切地说：一个平面图对于建筑师而言，平面中所呈现的点、线、面本身，是一个能够让人感觉到纵深的空间世界与图形，而二维的平面图只不过是认知三维世界的一个转换。

在建筑师的眼里，画面中的"形"本身，拥有着多意的存在。同时，对于平面图而言，也不像一般意义上的绘画那样具有透视层面的进深感（这种进深感源自文艺复兴以来的透视法则）。但是这些没有进深感的平面图本身事实上能够也可以视为：具有空间性和身体性意义的空间图形。如果从这样的理解方式再去观察绘画本身，不难发现：作为界定绘画本身画面的四角边框，其实是对"世界"片段疆域的裁剪，四角边框本身所构成的，是通向另一个"世界"领域的窗口。对于绘画而言，四角边框所截取的实际上还是"客观对象"（古典绘画）及"无客观对象""主观对象"（现代绘画）世界的场所与范围。这种绘画世界中"框景"的思考其实与建筑在大自然中以"框景"的举动来界定"建筑场地"范围的行为本质上如出一辙。沿着这样的理解，接下来我们从20世纪的现代绘画中选取35个实例，从绘画"框景"与建筑界定场地相一致的视点出发，寻求绘画与建筑二者在空间层面上的关联。

As for painting, relations among dots, lines, and planes, may be operated only as a pattern formation. Yet in the realm of architecture, dots, lines, and planes are a series of signs constituting space. When architects face the pattern design seemingly made up of dots, lines, and planes, they actually read a series of contents with space and function meanings, and also building indication. For example, "dots" may refers to pillars, "lines" refers to walls, and "planes" refers to roof… Therefore, when architects face the "patterns" made up of dots, lines, and planes, they instantly extend three-dimensional understanding. To be exact, dots, lines, and planes in a plan are space world and figure enabling depth to architects, while the two-dimensional plan is merely a transition to three-dimensional world cognition.

From architects' point of view, the "form" in the painting has a presence in multiple meanings. At the meantime, the plan is not like the traditional painting featured with perspective depth either (derived from perspective rules since Renaissance). Yet, these plain plans can actually be regarded as three-dimensional patterns with spatial and physical significance. Following similar thinking approach, it's not hard to find that painting's frame is in fact the crop of "world". What's defined by the frame, is a window opening towards another "world". As for painting, quadrangular frame also defines the worlds' realm and territory of "objective object" (classical painting), "non-objective object", and "subjective object" (modern painting). This "framed view" in painting is just like buildings framing views in nature to define "architecture site". Along this understanding, we then selected 35 modern painting samples in the 20th Century. Based on the viewpoint that painting "framing view" corresponds with architecture defining site, we intend to seek correlation between painting and architecture at space level.

左图：《有淡色块的圆形构图》绘画的空间关系图
Left: *Oval Composition with Light Colors*' spatial diagram

绘画与建筑关系的思考
Thoughts on correlation between painting and architecture

　　绘画所呈现的世界是画家大脑中所想要表达的一种意识空间，建筑师在进行建筑平面设计时，同样也是依据意识空间对空间进行一个划分的过程。无论从平面或立面的角度出发，绘画与建筑实际上都在通过线与线之间的关系对空间的构成进行指向，而绘画与建筑这两者在这个层面上，事实上已经获得了在本质上的统一。

The world presented by painting is an ideological space in painter's head. When architects do the graphic design, they also go through the similar space devision process based on ideological space. Whether starting off with plan or facade, painting and architecture are basically using relation between the lines to indicate space combination. Actually painting and architecture have achieved unity in nature.

绘画空间与建筑空间之间的关系
Correlation between painting and architecture spaces

之所以提及"绘画与建筑"这个话题,是因为绘画与建筑在本质上有着紧密的关联。建筑师在学习建筑设计之前,通常要先进行绘画练习,如速写、素描等,这似乎与绘画的学习过程相似。在学习了速写、素描等传统绘画形式的基础上,建筑学的教学还要转化为横、竖线条等制图方面的练习,这种制图的练习是一种空间表述的练习,从这时开始,似乎渐渐与绘画产生了分离。

传统意义上的绘画多是单纯地对"风景"等"客观对象"(古典绘画)具象的对象物进行表现与描绘,而建筑师的练习是一种点、线、面的制图,这两者虽都是在画,但之间存在着很大的差异。在建筑设计中,建筑师所画的草图,本质上就是在画线条,一根线、两根线……而这些线条与传统绘画中线条的意义实际上有所不同。在传统的对于"客观对象"(古典绘画)进行表现的绘画中,比如素描,通常主要进行一系列"具象的造型"的表现,但建筑师所画的每一根线条中,却包含着空间的意味,而这个空间的意味与建筑师大脑中的空间感是相对应的。又譬如,建筑师所画的几条线之间的关系,可以对应为几道墙之

The reason why we mention the topic of "Painting and Architecture", is because they're closely linked in essence. Before architects learn architecture design, they usually conduct painting exercises such as sketches and so on, very much the same as painting learning process. Having acquired traditional painting sketches skills, architecture students are required to do horizontal and vertical lines drawing practice, which is for spatial expression. From this time on, architecture seems to depart from painting.

Traditional painting is more likely to simply describe and render objective objects including "views" and concrete objects, while architects' exercise is drawing dots, lines, and planes. Although both are painting, they're significantly different. As in architecture design, the sketches by architects are actually line drawings, one line, two lines, and so on… Nevertheless, these lines have different meanings than in traditional paintings. For example, sketches in traditional paintings themed "objective objects", are to present a series of "concrete forms"; yet every line drawn by architects contains space meanings corresponding to the sense of space in architects' minds. For another example, relationship between the lines by architects may echo with spaces between a few walls. A horizontal line and two vertical lines in a plan can be

间的空间关系。建筑师同样是画一条横线、两条竖线，当我们将其置于平面中观察时，它可能就被视为一个平面图，横与竖的线条在其中起到的或许是分隔空间的作用；而当我们将其看作立面图或者剖面图时，横与竖的线条又可以表达一个屋顶与楼板和地面之间形成的空间关系。

由此，建筑师的绘与画和传统绘画中"风景"的绘画本质上是根本不同的，而这个不同，其实也是古典建筑和现代建筑之间存在着的一个重要不同。古典建筑虽然具有空间，也在满足使用，但其更主要是在表现其丰富的装饰性。

20世纪之后，物理学的新发现，相对论以及所谓关于时间、空间理论的出现，特别是伴随着摄影的出现与普及，通过绘画将某一刻凝固瞬间的意义失去了。相机可以瞬间将某种场景拍摄记录下来，于是人们对绘画究竟该表现什么提出了质疑和思考。

在这个时期，出现了一个与建筑新观念的发展密切相关、在现代绘画发展过程中起重要作用的"立体主义"流派，其艺术实践旨在探讨能否将时间的成分加入到绘画中。

seen as space partitions, and also can present the space formed by roof, floors and ground in facade or section.

Thus, drawing by architects and painting by traditional landscape painters are fundamentally different. And this difference also exist between classical architecture and modern architecture. Although classical architecture provides plentiful space meeting functional requirements, its rich decorative characters are more critical.

Entering into 20th Century, along with new Physics discoveries, Relativity theory and so-called time-space theories, especially invention and popularisation of photography, the significance of freezing the specific moment by painting is completely lost. Cameras can instantly shoot and record some scenario, so people begin to question what painting should present in the new era.

During this period, Cubism was born. It's closely related to architecture's new concept development, meanwhile playing an important role in modern painting advancement. Its artistic practice aims to explore the possibility of adding time element into painting.

Early Cubism was derived from Paul Cézanne's art

早期立体主义是从塞尚的艺术观念中发展而来的，塞尚并不是在对风景进行一个具象的描绘，他认为，世界是由几何学形体构成的，比如树是由圆锥或三角形构成的，这个世界总是可以由一些几何体表达出来。立体主义在塞尚的基础上，将客观的物体进行几何学处理。不过在那个时代，在从具象到非具象的转化过程中，画家自身的理解仍然存在一些偏差，有的画家力图将这个转化后的非具象物呈现出某种矿物质的形态。如布拉克的绘画《画了节拍器的静物》中，画面呈现出一种如同矿石的组合状态，这与20世纪初期许多艺术家以矿石、草木、枯萎的树干等为描绘的对象物的情绪是一致的。

立体主义的作品中还有一个原则就是在描绘一个物体时，不再只从一个固定角度去观察，而是将从不同角度观察的结果呈现在同一画面中，强调其"共时性"，如毕加索以人脸为对象的创作，运用格式塔心理学，对从不同角度观察到的脸，进行一种视觉上与知觉层面的游戏。

立体主义的发展分为三个阶段，第一阶段是早期立体主义，画面的表现较为具象和躁动。换句话说，

concept. Paul Cézanne didn't give concrete depiction of the landscape, instead, he thought the world is made up of geometrical forms, such as the tree is cone or triangle, and the whole world can be presented by a number of geometrical blocks. On the basis of Paul Cézanne's achievement, Cubism did geometrical transition on the objective objects. However, in that era, there were still some deviations in some painters' understanding of concrete-to-abstract transition. Some sought after mineral form for the abstract object after conversion. In Georges Braque's Painting Still Life with Metronome, its shows a combination of ore-like state, which went with the trend of many artists in the early 20th Century depicting objects such as ores, vegetation, and withered trunks.

There is another rule in Cubism works - "synchronic" is emphasised. Painters didn't only observe the object from one fixed angle, but also presented the different angles in one painting. Picasso's face drawings are well known for using Gestalt psychology, playing a visual and perceptive game with faces from various perspectives.

There are three stages of Cubism development. First stage is early Cubism, relatively concrete and agitated. In other words, audience can still tell the figure or object's

就是从画面中仍可以辨析到所描绘的人或对象物的形态；第二阶段是中期立体主义，此阶段的画面相比前一阶段规整了许多，但仍存在一些斜线，所描绘的人或对象物的形态也不那么容易辨析；而第三阶段是所谓的后期立体主义，这一阶段实际上是奥赞方和柯布西耶等人所主张的纯粹主义，此时期的画面变得十分几何化，通过计算，依据数的关系来描绘对象物。

　　谈到数这个层面，似乎所有的立体主义的绘画，即便是第三阶段的后期立体主义，其实其绘画本身实际上还是不够纯粹，因为不论从哪个层面来看，那些立体主义绘画中实际上还是能够看到"像"与变形了的"像"的存在。

　　绘画中的最纯粹者当属俄国画家马列维奇，他的至上主义绘画将绘画中的一切"像"加以涂抹，将所有一切转化为纯粹的方、圆等几何形体，并以涂抹掉"圣像"的画中人物而宣告"无对象"绘画即："无客观对象"＝"主观对象"（现代绘画）的开始。尽管其部分思想还是延续着塞尚的主张，但是绘画本身"无客观对象"的主张，迈出了连塞尚都难以突破的"对象物变形"的疆域。

form from the painting. Second stage is mid-Cubism, more regular compared with last stage, but still some slashes; the shape of figure or object is not that recognisable. As for the third stage, post-Cubism is actually Purism claimed by Ozenfant and Le Corbusier. The image becomes very geometric, object's proportion is depicted by calculation.

Speaking of numbers, it seems all Cubism paintings are not pure enough, even the third stage post-Cubism. Because no matter from which level, those Cubism paintings were still able to show the "image" and transformed "image".

The purist painting is undoubtedly by Russian painter Kasimier Severinovich Malevich. His suprematism works turn all "images" into pure geometric blocks like cube, sphere, etc.. He even announces "non-object" painting is namely the start of "non-objective object", "subjective object" (modern painting) by erasing the figures in "ikon" painting. Although he ideologically extends Paul Cézanne's ideas, the declaration of painting "non-objective object" makes a breakthrough of "object transformation" territory that even Paul Cézanne couldn't achieve.

Dutch De Stijl artists Mondrian and Doesburg can be

与俄国画家马列维奇相媲美的是荷兰的风格派,蒙德里安、杜斯伯格等人的作品。其强调并追求一种宇宙中基本的数的法则原理,力图表现宇宙空间的规律性与现实之间的对应关系,在哲学上,风格派紧追毕达哥拉斯的思想,认为"数是万物之源"。

绘画作品一旦与"数"有关,绘画本身与建筑师的操作产生关联似乎变得"顺理成章"。因为建筑的操作恰恰是在进行着一系列数的操作,如建筑的开间、进深、层高。而一旦以"数"的眼睛去看那些立体主义、至上主义绘画及风格派绘画中点线面本身所呈现的数理关系,瞬间地,这一系列绘画中所描绘的画面以及无意识间在画面中呈现的数理与几何形态的关系,居然瞬间开始已经可以并且能够与建筑空间产生关联。

从某种意义上讲,建筑本身就是以"线"为"墙"、以"点"为"柱"的一个空间表达方式,据此,如果一个绘画中的线、面之间表达的就是一种平面上的空间关系,那就十分容易地将绘画中的点、线、面转化成一个具象的空间,进而演化为一个建筑空间。

compared with Kasimier Severinovich Malevich. Their works emphasise and pursue a fundamental universal principle of numbers, keen to present the correlation between regularity and reality in the universe space. Philosophically speaking, De Stijl is Pythagoras's follower on the idea that "number is the origin of everything."

Once the painting is connected with "numbers", correlation between painting per-se and architectural practice seems logical. Because architectural practice is precisely the operation of a series of numbers, such building's width, depth, and height. As soon as we look at the mathematical relationship presented by dots, lines, and planes in Cubism, Suprematism Painting, and De Stijl Painting through the eyes of "numbers", the pictures depicted by the series of paintings and following mathematical and geometrical relationship unconsciously shown in the paintings, momentarily begin to associate with architecture space.

In some sense, architecture itself is a space expression using "lines" as "walls" and "points" as "columns". Accordingly, if the lines and planes in a painting present a space relationship on the plane, it's very easy to transit dots, lines, and planes into a concrete space, and therefore evolve into an architecture space.

沿着上面的思路再来重新审视20世纪的绘画，其实很多绘画作品既不是平面性的绘画，也不是透视的绘画，而是几个层面叠加所呈现的前面所谈到的"共时性"状态，尽管古典绘画也具有多个空间层次，但那是为了追求现实感的远景、中景、近景。而立体派的绘画不仅包含远景、中景、近景，还具有背面、侧面、上顶、下底……将微小的变化不断地加以叠加以表达时间的共时性。这种将不同的场景叠加、透明化并置于同一画面的表达观念其实与强调对建筑空间的观察是观察者在建筑空间中游走所产生叠加的结果的理解相互一致。

本着这样的思考与理解，在这本小册子中，我们将视点落在20世纪的现代绘画，从中抽取布拉克、马列维奇、蒙德里安、列克、杜斯堡、格拉尔纳、米罗、马蒂斯、克利、康定斯基等人部分具有代表性的绘画作品，通过对其进行空间化、立体化处理以及转化实例的展示与试做，以明示绘画与建筑的关系及从绘画转换到建筑空间的可能性。

Using the above-said thinking to re-examine the 20th Century painting, many of them are neither planar nor perspective, but "synchronic" juxtaposed by several layers. Although traditional painting also has multiple spatial levels, it's merely in pursuit of a sense of reality-long-range, medium, and close range views. While Cubism paintings include not only these views, but also the back, profile, top, and bottom… tiny changes are constantly superimposed to express synchronic state. This expression concept of different scenes superposition, transparentizing, and placed in the same picture, is in fact consistent with the architecture space observation being the superposition of architecture promenade.

In line with such thinking and understanding, in this book we will base on 20th Century modern painting, choose representative paintings of Georges Braque, Kasimier Severinovich Malevich, Mondrian, Leck, Theo van Doesburg, Glanrner, Miro, Matisse, Klee, and Kandinsky, showcase and test on spatial, three-dimensional processing and conversion, to present the relationship between painting and architecture, and transition possibility from painting to architecture space.

绘画中的空间与建筑中的空间
Spaces in painting and architecture

　　本章节针对32个绘画作品，进行绘画与空间关系上的试做，这些绘画的样本是从20世纪的现代绘画中抽取的。原本只是作为绘画来进行观察的这些作品，一旦以建筑师的空间视点（请注意：这一点非常重要，是建筑师的视点而不是画家的视点，更不是平面设计师二维眼睛的视点，而是建筑师三维眼睛的视点）来进行观察，绘画中的点、线、面的关系便瞬间地产生空间层次和空间开闭关系。一旦这种空间感得以呈现，瞬间地，观察者便会将自己的身体投入到画面中，投射到由点、线、面所呈现的空间关系中，进而能够从平面的绘画中察觉、感受并拥有空间意义上的感受和关联。这一切实际上便是绘画与建筑之间产生关联的开始。

This chapter is the experiment on relation between paintings and space using 32 painting samples, selected from 20th Century modern paintings. Once we use architects spatial viewpoint (NB this is very important, it's architects' three-dimensional perspective instead of painters', nor graphic designers' planar viewpoint) to observe these paintings, relationship among dots, lines, and planes instantly generate spatial layers and on/off relation. If such a sense of space is presented, the observer would instantly devote his body into the picture, project onto the spatial relationship shown by dots, lines, and planes. Furthermore, the observer is able to detect from planar painting, feel and own the spatial affections and association, all of which are the beginning where painting connects with architecture.

从绘画空间到建筑空间表达的32个实例
32 Samples from painting to architecture spatial expression

实例1《码头和海》
绘画的空间性表达
Sample 1 *Pier and Ocean*'s spatial expression

《码头和海》绘画为荷兰画家彼埃·蒙德里安于1914年所绘。在这幅绘画中，蒙德里安将水面上晶莹跳动的光及其中所蕴含的情绪，通过在一个椭圆形区域内散落分布的十字形的抽象形式进行表达。我们从平面的视角读解这幅绘画，获取空间中的点、线关系，形成《码头和海》绘画的空间组成图（图1-1），进而对其进行空间立体化处理（给予进深），获得《码头和海》绘画的抽象空间关系图（图1-2、图1-3）。

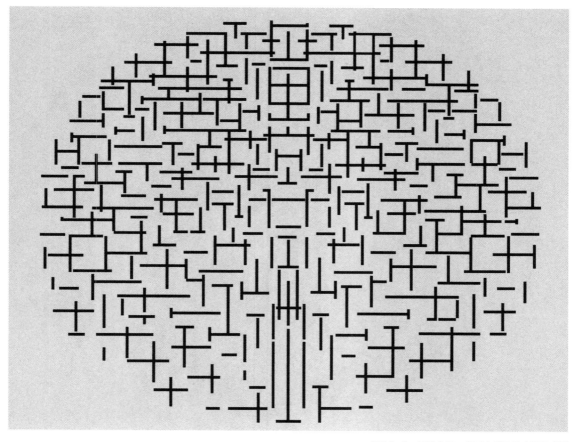

图1-1 从《码头和海》绘画中抽取的空间组成图
Fig. 1-1 Spatial composition diagram extracted from *Pier and Ocean*

Pier and Ocean was completed by Dutch painter Piet Mondrian in 1914. In this painting, by means of very abstract expression form of crosses scattered in an oval area, Mondrian presented glittery light above the crystal clear water and inherent sentiment. We read this piece of painting from the perspective of the plane, get the dots-lines relation, and form the spatial composition diagram of *Pier and Ocean* (Fig. 1-1). Then we add depth to make it three-dimensional, therefore obtain *Pier and Ocean*'s abstract spatial diagrams (Fig. 1-2, Fig. 1-3).

图1-2 《码头和海》绘画的空间关系图1
Fig. 1-2 *Pier and Ocean*'s spatial diagram 1

图1-3 《码头和海》绘画的空间关系图2
Fig. 1-3 *Pier and Ocean*'s spatial diagram 2

实例2《开花的苹果树》
绘画的空间性表达
Sample 2 *The Flowering Apple Tree*'s spatial expression

《开花的苹果树》绘画为荷兰画家彼埃·蒙德里安于1912年所绘。在这幅绘画中，蒙德里安通过一段段在水平和垂直方向上受到控制的黑色抽象弧线，表现从现实中具象的树木中抽象出的结构本质。我们从平面的视角读解这幅绘画，获取空间中的点、线关系，形成《开花的苹果树》绘画的空间组成图（图2-1），进而再对其进行空间立体化处理（给予进深），获得《开花的苹果树》绘画的抽象空间关系图（图2-2、图2-3）。

图2-1 从《开花的苹果树》绘画中抽取的空间组成图
Fig. 2-1 Spatial composition diagram extracted from *The Flowering Apple Tree*

The Flowering Apple Tree was painted by Piet Mondrian in 1912. In this painting, by means of some black abstract arcs controlled in horizontal and vertical directions, Mondrian tried to present the structural nature abstracted from the real trees. We read this piece of painting from the perspective of the plane, get the dots-lines relation, and form the spatial composition diagram of *The Flowering Apple Tree* (Fig. 2-1). Then we add depth to make it three-dimensional, therefore obtain *The Flowering Apple Tree*'s abstract spatial diagrams (Fig. 2-2, Fig. 2-3).

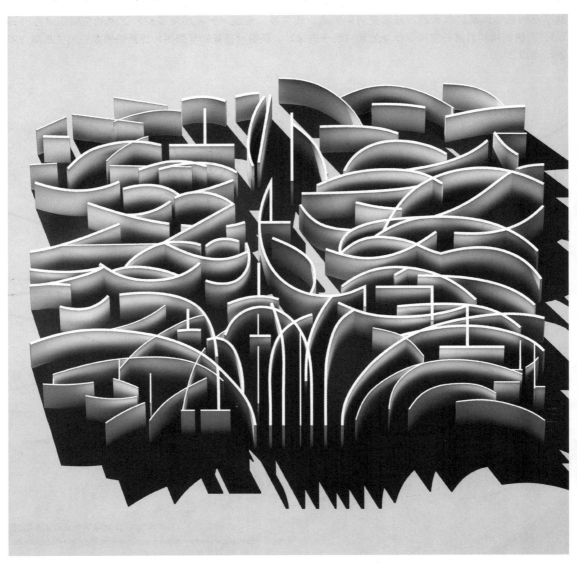

图2-2 《开花的苹果树》绘画的空间关系图1
Fig. 2-2 *The Flowering Apple Tree*'s spatial diagram 1

图2-3 《开花的苹果树》绘画的空间关系图2
Fig. 2-3 *The Flowering Apple Tree*'s spatial diagram 2

实例3《有线条的构图（黑和白的构图）》
绘画的空间性表达
Sample 3 *Composition with Lines (Composition in Black and White)*'s spatial expression

《有线条的构图（黑和白的构图）》绘画为荷兰画家彼埃·蒙德里安于1917年所绘。在这幅绘画中，蒙德里安在一个圆形区域内，描绘了一组散落的水平、垂直黑色短粗直线。我们从平面的视角读解这幅绘画，获取空间中的点、线关系，形成《有线条的构图》绘画的空间组成图（图3-1），进而再对其进行空间

Composition with Lines (Composition in Black and White) was painted by Piet Mondrian in 1917. In this painting, Mondrian draws a group of scattered horizontal and vertical black thick lines within a circular area. We read this piece of painting from the perspective of the plane, get the dots-lines relation, and form the spatial composition diagram (Fig.

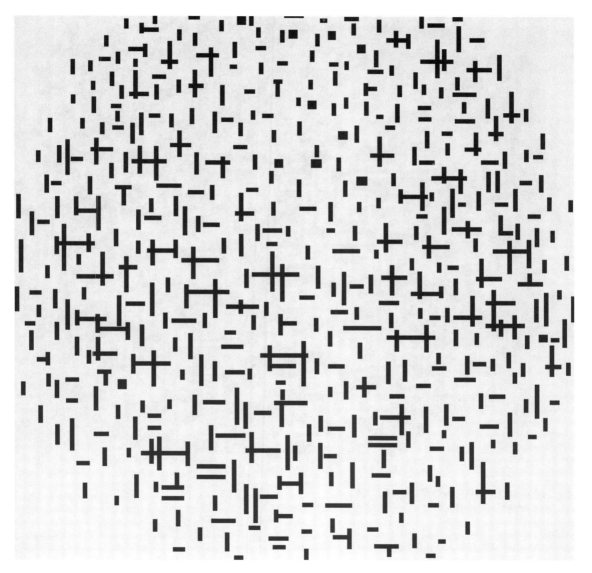

图3-1 从《有线条的构图（黑和白的构图）》绘画中抽取的空间组成图

Fig. 3-1 Spatial composition diagram extracted from *Composition with Lines (Composition in Black and White)*

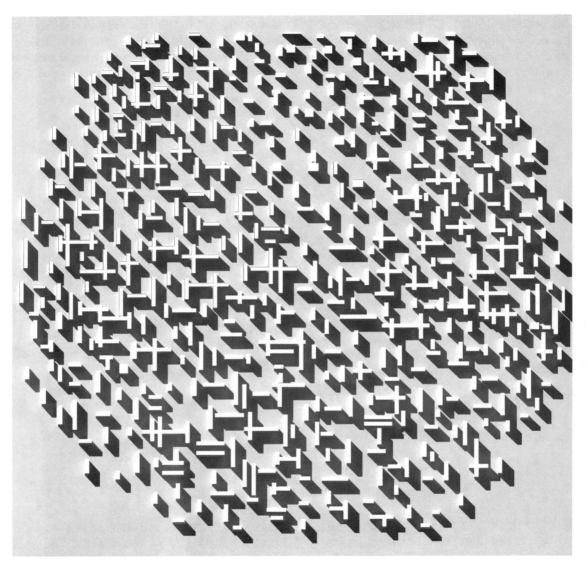

图3-2 《有线条的构图（黑和白的构图）》绘画的空间关系图1
Fig. 3-2 *Composition with Lines (Composition in Black and White)*'s spatial diagram 1

立体化处理，对画面中的黑色短粗直线给予进深，转化为内部含有空间的体块，获得《有线条的构图》绘画的抽象空间关系图（图3-2、图3-3）。

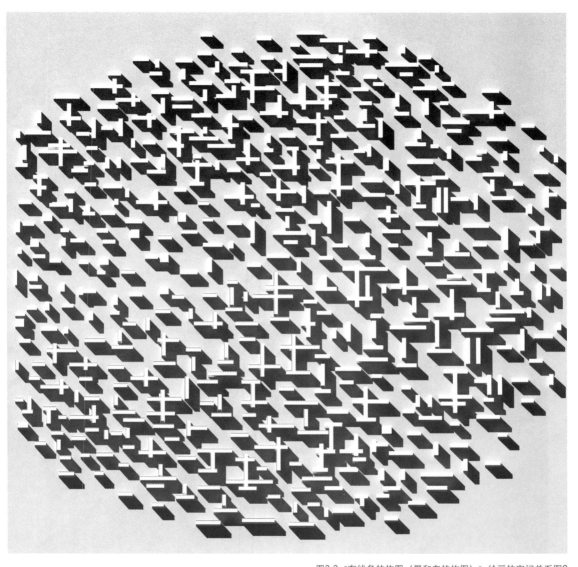

图3-3 《有线条的构图（黑和白的构图）》绘画的空间关系图2
Fig. 3-3 *Composition with Lines (Composition in Black and White)*'s spatial diagram 2

3-1). Then we add depth to the black thick short lines, turn it into blocks containing interior spaces, therefore obtain *Composition with Lines (Composition in Black and White)*'s abstract spatial diagrams (Fig. 3-2, Fig. 3-3).

实例4《有色块的构图B》
绘画的空间性表达
Sample 4 *Composition in Color B*'s spatial expression

《有色块的构图B》绘画为荷兰画家彼埃·蒙德里安于1917年所绘。这幅绘画由水平、垂直放置的黑色短粗结构线条和彩色矩形构成。我们从平面的视角读解这幅绘画,获取空间中的点、线关系,形成《有
Composition in Color B was painted by Piet Mondrian in 1917. This painting is composed of black short thick horizontal and vertical structural lines, and rectangles in colors. We read this piece of painting from the perspective of the plane,

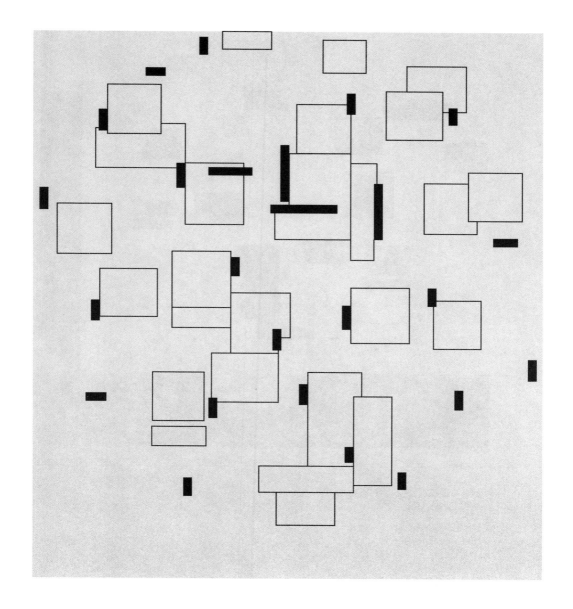

图4-1 从《有色块的构图B》绘画中抽取的空间组成图
Fig. 4-1 Spatial composition diagram extracted from *Composition in Color B*

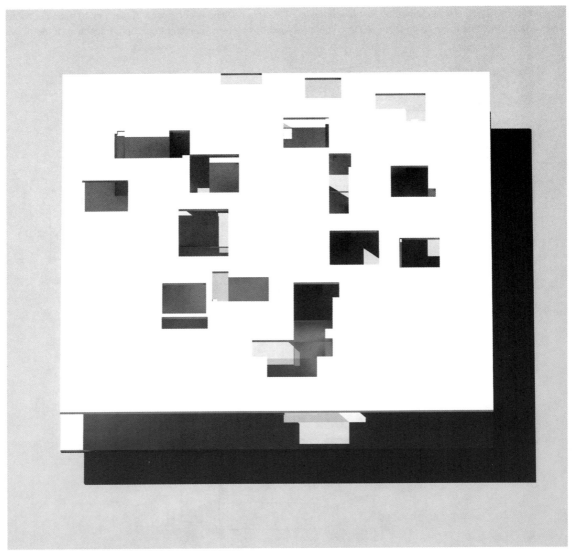

图4-2 《有色块的构图B》绘画的空间关系图1
Fig. 4-2 Composition in Color B's spatial diagram 1

色块的构图B》绘画的空间组成图（图4-1），进而再对其进行空间立体化处理，以画框为界面的外部边界、彩色矩形的轮廓为内部边界，对这两组边界所挤压出的界面以及画面中黑色短粗结构线条给予进深，获得《有色块的构图B》绘画的抽象空间关系图（图4-2、图4-3）。

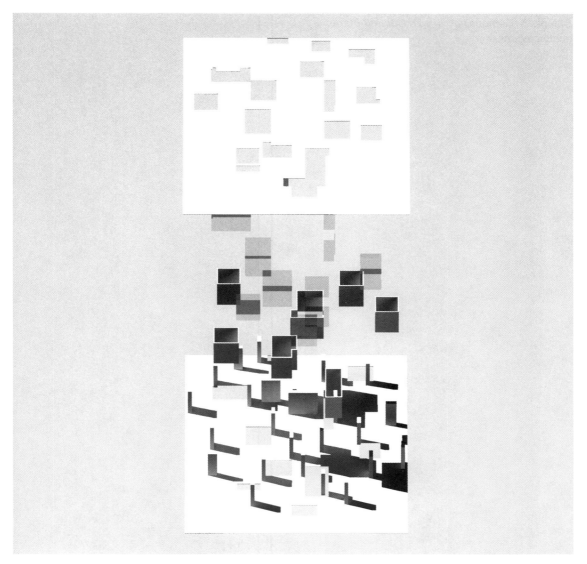

图4-3 《有色块的构图B》绘画的空间关系图2
Fig. 4-3 *Composition in Color B*'s spatial diagram 2

get the dots-lines relation, and form the spatial composition diagram of *Composition in Color B* (Fig. 4-1). Then we make it three-dimensional. Use the frame as external border and coloured rectangle as internal border, add depth to in-betweens and the black short thick structural lines, therefore obtain *Composition in Color B*' s abstract spatial diagrams (Fig. 4-2, Fig. 4-3).

实例5《色彩平衡构成三号》
绘画的空间性表达
Sample 5 *Composition III with Colored Planes*' spatial expression

《色彩平衡构成三号》绘画为荷兰画家彼埃·蒙德里安于1917年所绘。在这幅绘画中,蒙德里安绘制了一组受到正交网格控制的彩色矩形。我们从平面的视角读解这幅绘画,获取空间中的点、线关系,形成《色彩平衡构成三号》绘画的空间关系图(图5-1),并对其进行两种空间立体化处理:其一,对画面中的彩色矩形给予进深,转化为内部含有空间的体块,获得《色彩平衡构成三号》绘画的第一种抽象空间组成

图5-1 从《色彩平衡构成三号》绘画中抽取的空间组成图
Fig. 5-1 Spatial composition diagram extracted from *Composition III with Colored Planes*

Composition III with Colored Planes was painted by Piet Mondrian in 1917. In this painting, Mondrian draws a set of coloured rectangles in orthogonal grid. We read this piece of painting from the perspective of the plane, get the dots-lines relation, and form the spatial composition diagram of *Composition III with Colored Planes* (Fig. 5-1). Then we make it three-dimensional in two approaches. One, we add depth to coloured rectangles, turn it into blocks containing interior spaces, therefore obtain first type of *Composition III with Colored Planes*' abstract spatial diagrams (Fig. 5-2,

图5-2 《色彩平衡构成三号》绘画的空间关系图1

Fig. 5-2 *Composition III with Colored Planes*' spatial diagram 1

图（图5-2、图5-3）；其二，对第一种读解方式进行反读，即仅仅保留绘画的边框，将其中的矩形去掉，获得具有廊子的空间感的第二种抽象空间关系图（图5-4、图5-5）。

图5-3 《色彩平衡构成三号》绘画的空间关系图2
Fig. 5-3 *Composition III with Colored Planes*' spatial diagram 2

Fig. 5-3). Two, we do the opposite, i.e. just keep the painting's frame and remove all the rectangles, therefore obtain second type of abstract porch-like spatial diagrams (Fig. 5-4, Fig. 5-5).

图5-4 《色彩平衡构成三号》绘画的空间关系图3

Fig. 5-4 *Composition III with Colored Planes'* spatial diagram 3

图5-5 《色彩平衡构成三号》绘画的空间关系图4
Fig. 5-5 *Composition III with Colored Planes*' spatial diagram 4

实例6《构成4号（离开工厂）》
绘画的空间性表达
Sample 6 *Composition No.4 (Leaving the Factory)*'s spatial expression

《构成4号（离开工厂）》绘画为荷兰画家巴特·凡·德·列克于1917年所绘。在这幅绘画中，列克绘制了受正交网格控制的彩色和黑色矩形。我们从平面的视角读解这幅绘画，获取空间中的点、线关系，形成《构成4号（离开工厂）》绘画的空间组成图（图6-1），并对其进行两种空间立体化处理：其一，对 Composition No.4 (Leaving the Factory) was completed by Dutch painter Bart van der Leck in 1917. In this painting, Leck draws coloured and black rectangles in orthogonal grid. We read this piece of painting from the perspective of the plane, get the dots-lines relation, and form the spatial composition diagram of Composition No.4 (Leaving the Factory)

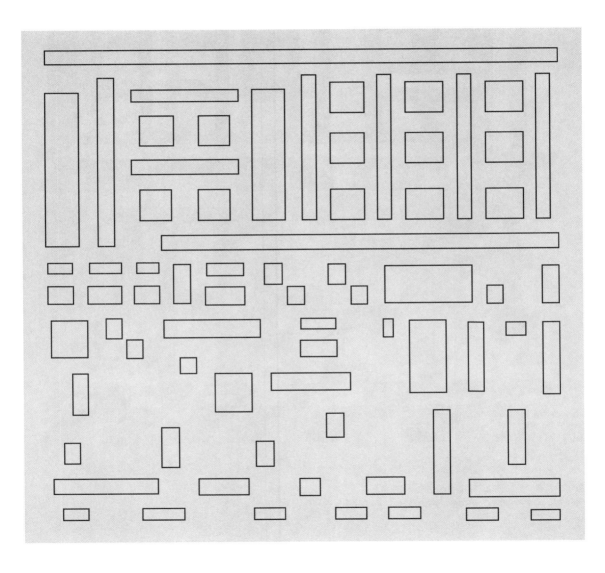

图6-1 从《构成4号（离开工厂）》绘画中抽取的空间组成图

Fig. 6-1 Spatial composition diagram extracted from Composition No.4 (Leaving the Factory)

图6-2《构成4号(离开工厂)》绘画的空间关系图1

Fig. 6-2 Composition No.4 (Leaving the Factory)'s spatial diagram 1

画面中的矩形给予进深,转化为内部含有空间的体块,获得《构成4号(离开工厂)》绘画的第一种抽象空间关系图(图6-2、图6-3);其二,对第一种读解方式进行反读,即仅仅保留绘画的边框,将其中的矩形去掉,获得具有廊子的空间感的第二种抽象空间关系图(图6-4、图6-5)。

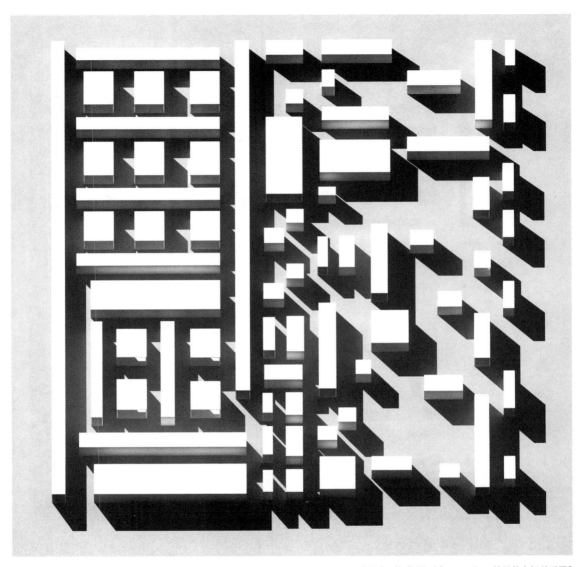

图6-3 《构成4号（离开工厂）》绘画的空间关系图2
Fig. 6-3 Composition No.4 (Leaving the Factory)'s spatial diagram 2

(Fig. 6-1). Then we make it three-dimensional in two approaches. One, we add depth to rectangles, turn it into blocks containing interior spaces, therefore obtain first type of Composition No.4 (Leaving the Factory) (Fig. 6-2, Fig. 6-3). Two, we do the opposite, i.e. just keep the painting's frame and remove all the rectangles, therefore obtain second type of abstract porch-like spatial diagrams (Fig. 6-4, Fig. 6-5).

图6-4 《构成4号（离开工厂）》绘画的空间关系图3
Fig. 6-4 Composition No.4 (Leaving the Factory)'s spatial diagram 3

图6-5《构成4号（离开工厂）》绘画的空间关系图4
Fig. 6-5 Composition No.4 (Leaving the Factory)'s spatial diagram 4

实例7《纽约市Ⅰ》
绘画的空间性表达
Sample 7 *New York City I*'s spatial expression

《纽约市Ⅰ》为荷兰画家彼埃·蒙德里安于1941年绘制的风格派绘画。在这幅绘画中，蒙德里安将不同颜色的正交网格进行编织，形成一种新的构成。我们从平面的视角读解这幅绘画，获取空间中的点、线

New York City I was completed by Dutch painter Piet Mondrian in 1941. In this painting, Mondrian interweaves orthogonal grids of different colors to form a new combination. We read this piece of painting from the perspective of

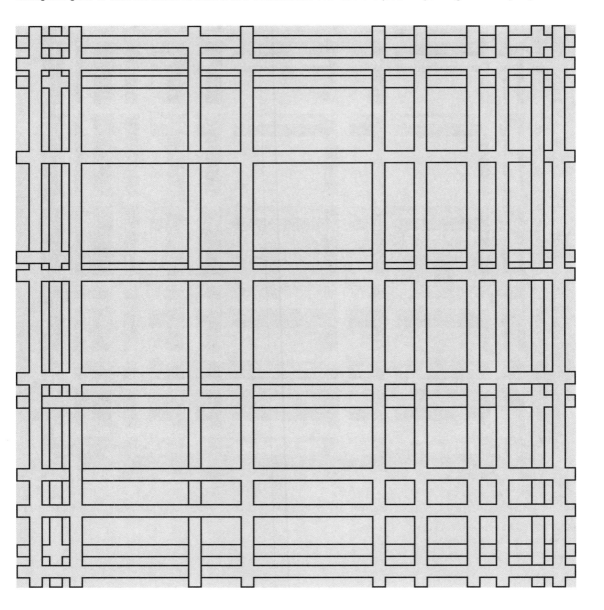

图7-1 从《纽约市Ⅰ》绘画中抽取的空间组成图

Fig. 7-1 Spatial composition diagram extracted from *New York City I*

41

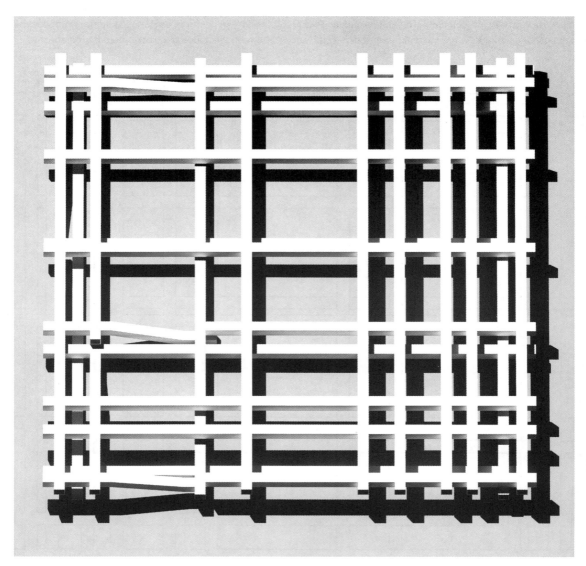

图7-2 《纽约市Ⅰ》绘画的空间关系图1

Fig. 7-2 *New York City I*'s spatial diagram 1

关系,形成《纽约市Ⅰ》绘画的空间组成图(图7-1),进而再对其进行空间立体化处理,将正交网格按其不同的颜色分为三层相互编织的管状空间,获得《纽约市Ⅰ》绘画的抽象空间关系图(图7-2~图7-6)。

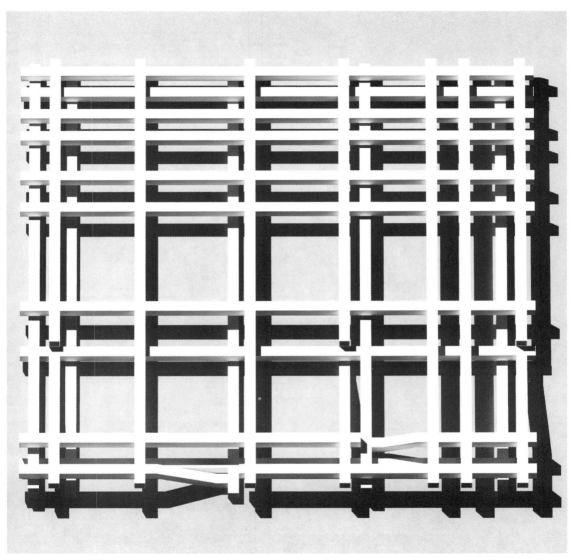

图7-3 《纽约市Ⅰ》绘画的空间关系图2
Fig. 7-3 *New York City I*'s spatial diagram 2

the plane, get the dots-lines relation, and form the spatial composition diagram of *New York City I* (Fig. 7-1). Then we make it three-dimensional, divide the orthogonal grids into three interwoven tubular spaces by their colors, therefore obtain *New York City I*'s abstract spatial diagrams (Fig. 7-2—Fig. 7-6).

图7-4 《纽约市Ⅰ》绘画的空间关系图3
Fig. 7-4 New York City I's spatial diagram 3

45

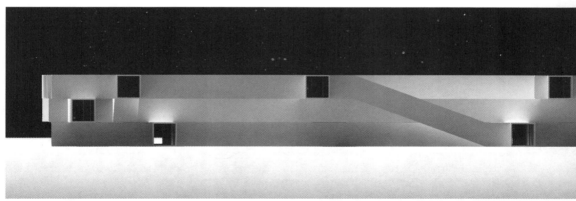

图7-5 《纽约市Ⅰ》绘画的空间关系图4
Fig. 7-5 *New York City I*'s spatial diagram 4

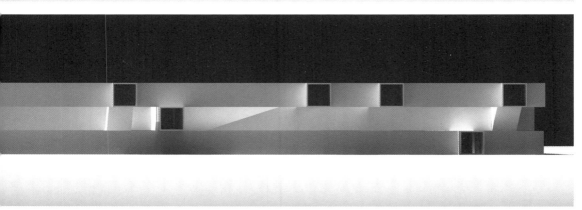

图7-6《纽约市Ⅰ》绘画的空间关系图5
Fig. 7-6 *New York City I*'s spatial diagram 5

实例8《俄罗斯舞蹈的韵律》
绘画的空间性表达
Sample 8 *Rhythms of a Russian Dance*'s spatial expression

《俄罗斯舞蹈的韵律》为荷兰画家特奥·凡·杜斯堡于1918年绘制的风格派绘画。这幅绘画由水平、垂直放置的彩色、黑色短粗结构线条构成，并在其内部围合出一个浅色的部分。我们从平面的视角读解这

Rhythms of a Russian Dance was completed by Dutch painter Theo van Doesburg in 1918. The painting is composed of horizontal and vertical coloured and black short thick structural lines, with a light-coloured part enclosed. We read

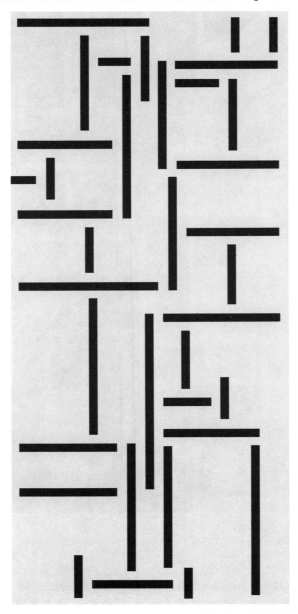

图8-1 从《俄罗斯舞蹈的韵律》绘画中抽取的空间组成图
Fig. 8-1 Spatial composition diagram extracted from *Rhythms of a Russian Dance*

图8-2 《俄罗斯舞蹈的韵律》绘画的空间关系图1
Fig. 8-2 Rhythms of a Russian Dance's spatial diagram 1

幅绘画，获取空间中的点、线关系，形成《俄罗斯舞蹈的韵律》绘画的空间组成图（图8-1），进而再对其进行空间立体化处理（给予进深），以结构线条为墙体、浅色部分为建筑室内空间，获得《俄罗斯舞蹈的韵律》绘画的抽象空间关系图（图8-2、图8-3）以及场景图（图8-4、图8-5）。

图8-3 《俄罗斯舞蹈的韵律》绘画的空间关系图2
Fig. 8-3 *Rhythms of a Russian Dance*'s spatial diagram 2

this piece of painting from the perspective of the plane, get the dots-lines relation, and form the spatial composition diagram of *Rhythms of a Russian Dance* (Fig. 8-1). Then we add depth to make it three-dimensional - use structural lines as walls and light-colored part as interior space, therefore obtain *Rhythms of a Russian Dance*'s abstract spatial diagrams (Fig. 8-2, Fig. 8-3) and space scenes (Fig. 8-4, Fig. 8-5).

图8-4《俄罗斯舞蹈的韵律》绘画的空间场景1
Fig. 8-4 *Space scenes of Rhythms of a Russian Dance* 1

图8-5 《俄罗斯舞蹈的韵律》绘画的空间场景2
Fig. 8-5 Space scenes of *Rhythms of a Russian Dance* 2

实例9 《圆形画 51》
绘画的空间性表达
Sample 9 *Tondo 51*'s spatial expression

《圆形画 51》为美国画家弗里茨·格拉尔纳基于1958年绘制的风格派绘画。在这幅绘画中，创作被控制在一个圆形的区域内，外围绘制了一圈具有一定宽度的、被进行多段切割的结构轮廓。在一套正交的

Tondo 51 was completed by American painter Fritz Glanrner in 1958. In this painting, the creation is limited within a circular area, surrounded by the structural outlines of a certain width in segments. Under the control of a set of orthogonal axes, the circular area is split by rectangle blocks in different colors and structural lines with a certain

图9-1 从《圆形画 51》绘画中抽取的空间组成图

Fig. 9-1 Spatial composition diagram extracted from *Tondo 51*

图9-2《圆形画 51》绘画的空间关系图1
Fig. 9-2 *Tondo 51*'s spatial diagram 1

轴线控制下,通过不同颜色的矩形色块和具有一定宽度的结构线条对这个圆形区域进行了分割。我们从平面的视角进行读解,获取空间中的点、线关系,形成《圆形画 51》绘画的空间组成图(图9-1),进而再对其进行空间立体化处理,获得《圆形画 51》绘画的抽象空间关系图(图9-2、图9-3)。

图9-3 《圆形画 51》绘画的空间关系图2
Fig. 9-3 *Tondo 51*'s spatial diagram 2

width. We read this piece of painting from the perspective of the plane, get the dots-lines relation, and form the spatial composition diagram of *Tondo 51* (Fig. 9-1). Then we add depth to make it three-dimensional, therefore obtain *Tondo 51*'s abstract spatial diagrams (Fig. 9-2, Fig. 9-3).

实例10《爱国庆祝会（自由单词绘画）》
绘画的空间性表达
Sample 10 *Interventionist Demonstration (Free Word Painting)*'s spatial expression

《爱国庆祝会（自由单词绘画）》是意大利未来主义画家卡洛·卡拉创作于1914年的纸板拼贴画。这幅绘画以各种字体不同的书报、杂志、乐谱等印刷品为材料，利用其中的"自由单词"组构画面。卡拉将

Interventionist Demonstration (Free Word Painting) was completed by Italian Futurist painter Carlo Carra in 1914. This cardboard collage uses materials including books, newspapers, magazines, and music score clippings in a variety of

图10-1 从《爱国庆祝会（自由单词绘画）》绘画中抽取的空间组成图
Fig. 10-1 Spatial composition diagram extracted from *Interventionist Demonstration (Free Word Painting)*

图10-2《爱国庆祝会(自由单词绘画)》绘画的空间俯视图
Fig. 10-2 *Interventionist Demonstration (Free Word Painting)*'s bird view

它们剪裁、拼贴,形成从中心向四周放射的构图。我们从平面的视角读解这幅绘画,获取空间中的点、线关系,形成《爱国庆祝会(自由单词绘画)》绘画的空间组成图(图10-1),进而再对其进行空间立体化处理(给予进深),获得《爱国庆祝会(自由单词绘画)》绘画的空间整体俯视图(图10-2)以及抽象空间关系图(图10-3、图10-4)。

图10-3 《爱国庆祝会（自由单词绘画）》绘画的空间关系图1
Fig. 10-3 *Interventionist Demonstration (Free Word Painting)*'s spatial diagram 1

fonts, builds up the picture by different kinds of "free word". Carlo Carra cuts and collages to form a radiation pattern. We read this piece of painting from the perspective of the plane, get the dots-lines relation, and form the spatial composition diagram of *Interventionist Demonstration (Free Word Painting)* (Fig. 10-1). Then we add depth to make it three-dimensional, therefore obtain bird view (Fig. 10-2) and abstract spatial diagram (Fig. 10-3, Fig. 10-4).

图10-4 从《爱国庆祝会（自由单词绘画）》绘画的空间关系图2

Fig. 10-4 *Interventionist Demonstration (Free Word Painting)*'s spatial diagram 2

实例11《男人的头》
绘画的空间性表达
Sample 11 *Man's Head*'s
spatial expression

《男人的头》绘画为西班牙画家杰昂·米罗于1932年所绘。在这幅绘画中，曲线将画面划分成几个色块。我们从平面的视角读解这幅绘画，获取空间中的点、线关系，形成《男人的头》绘画的空间组成

Man's Head was completed by Spanish painter Joan Miro in 1932. In this painting, the picture is divided into several colour blocks by curves, to present a figurative image of *Man's head*. We read this piece of painting from the

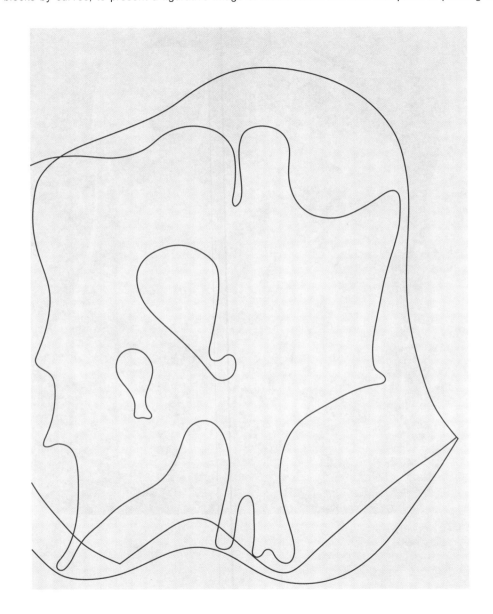

图11-1 从《男人的头》绘画中抽取的空间组成图
Fig. 11-1 Spatial composition diagram extracted from *Man's Head*

图11-2 《男人的头》绘画的空间关系图1
Fig. 11-2 *Man's Head*'s spatial diagram 1

图（图11-1），进而再通过对其图底关系的不同读解方式，对其进行多种空间立体化处理，获得《男人的头》绘画的一系列抽象空间关系图（图11-2～图11-5）。当我们将绘画中大面积的浅色部分视为院落，其

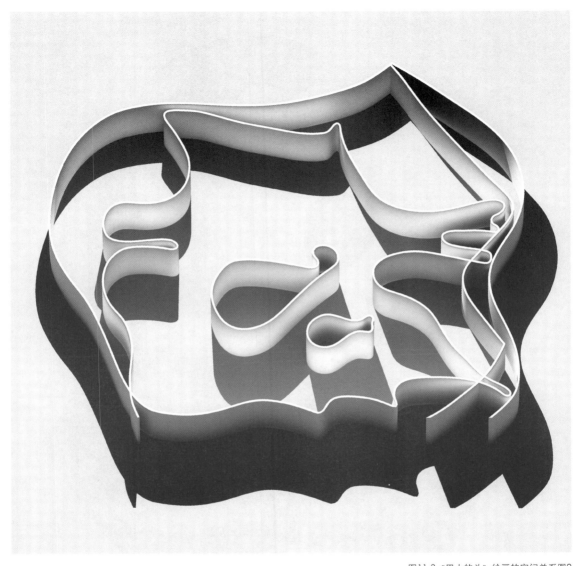

图11-3 《男人的头》绘画的空间关系图2
Fig. 11-3 *Man's Head*'s spatial diagram 2

perspective of the plane, get the dots-lines relation, and form the spatial composition diagram of *Man's Head* (Fig. 11-1). Then we make it three-dimensional through different interpretations of figure-ground relation, therefore obtain a series of *Man's Head*'s abstract spatial diagrams (Fig. 11-2 — Fig. 11-5). When we deem the large portion of light-

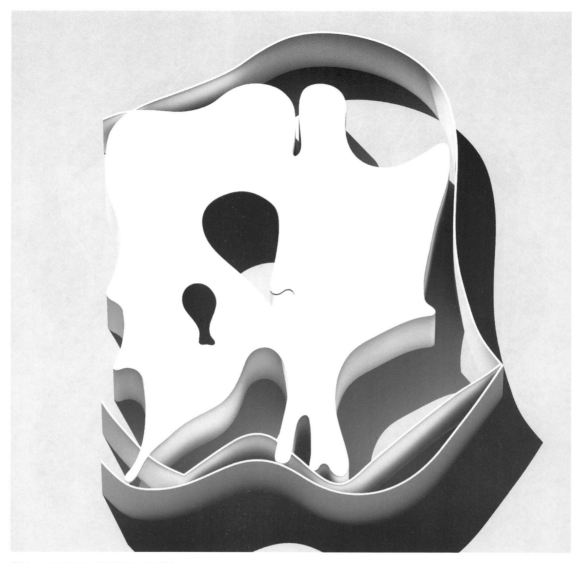

图11-4《男人的头》绘画的空间关系图3
Fig. 11-4 *Man's Head*'s spatial diagram 3

周围的深色部分视为廊道,位于画面中间位置的两个有机图形视为水榭时,会获得一个近似中国园林的空间构成(图11-6、图11-7),参考本人拙作《建筑与园林》。

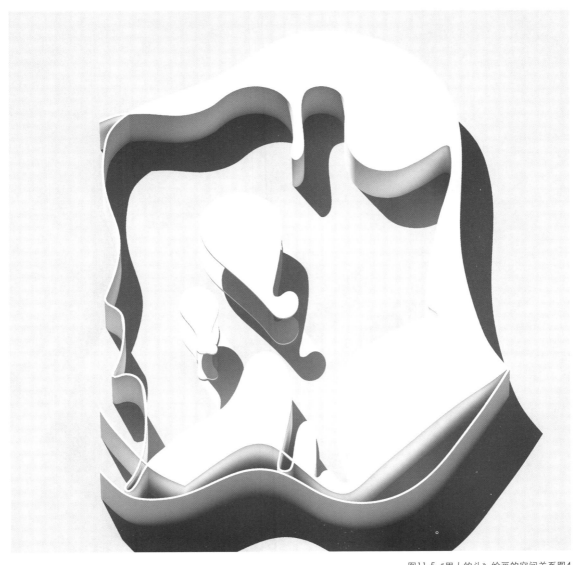

图11-5 《男人的头》绘画的空间关系图4
Fig. 11-5 *Man's Head*'s spatial diagram 4

coloured area as courtyard, surrounding dark area as corridors, and two organic graphics in the middle as waterside pavilions, we get some kind of Chinese garden consitution (Fig. 11-6, Fig. 11-7), with reference to my book *Architecture and Gardens*.

图11-6《男人的头》绘画的空间关系图5
Fig. 11-6 *Man's Head*'s spatial diagram 5

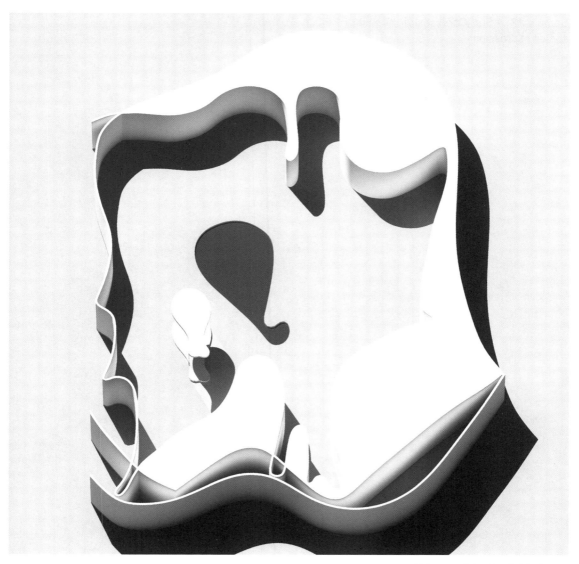

图11-7 《男人的头》绘画的空间关系图6
Fig. 11-7 Man's Head's spatial diagram 6

实例12《抚摸黑人妇女胸部的一颗星》
绘画的空间性表达
Sample 12 *A Star Caresses the Breast of a Black Woman*'s spatial expression

《抚摸黑人妇女胸部的一颗星》绘画为西班牙画家杰昂·米罗于1938年所绘。画面由自由的曲线、曲线形的文字和由曲线围合而成的色块构成，类似一个基地上不同曲线关系的罗列。螺旋形的线、三角形的面及体块，构成一种新的场域的概念。我们从平面的视角读解这幅绘画，获取空间中的点、线关系，形成《抚摸黑人妇女胸部的一颗星》绘画的空间组成图（图12-1），进而再对其进行空间立体化处理，获得《抚摸黑人妇女胸部的一颗星》绘画的抽象空间关系图（图12-2、图12-3）。

图12-1 从《抚摸黑人妇女胸部的一颗星》绘画中抽取的空间组成图
Fig. 12-1 Spatial composition diagram extracted from *A Star Caresses the Breast of a Black Woman*

A Star Caresses the Breast of a Black Woman was completed by Spanish painter Joan Miro in 1938. The picture is composed of free-style curves, curve-shaped texts and color blocks enclosed by curves. So it's basically a combinations of different curve relationship on on site. Spiral lines, triangular planes, and blocks, constitute a new concept of field. We read this piece of painting from the perspective of the plane, get the dots-lines relation, and form the spatial composition diagram of *A Star Caresses the Breast of a Black Woman* (Fig. 12-1). Then we make it three-dimensional, therefore obtain *A Star Caresses the Breast of a Black Woman*'s abstract spatial diagrams (Fig. 12-2, Fig. 12-3).

图12-2 《抚摸黑人妇女胸部的一颗星》绘画的空间关系图1
Fig. 12-2 *A Star Caresses the Breast of a Black Woman*'s spatial diagram 1

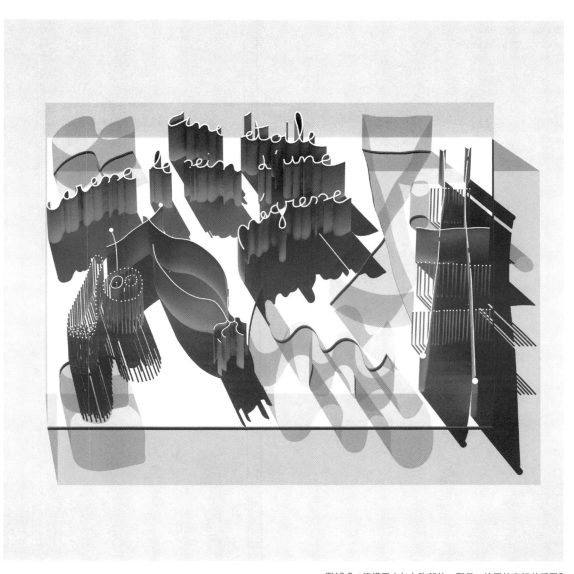

图12-3 《抚摸黑人妇女胸部的一颗星》绘画的空间关系图2
Fig. 12-3 *A Star Caresses the Breast of a Black Woman*'s spatial diagram 2

实例13《蓝色二号》
绘画的空间性表达
Sample 13 *Blue II*'s
spatial expression

《蓝色二号》绘画为西班牙画家杰昂·米罗于1961年所绘。米罗于蓝色的底色中绘制了一行沿水平排布的大小不一的不规则的黑色圆点,以及一条竖向放置的红色笔触。我们从平面的视角读解这幅绘画,获取空间中的点、线关系,形成《蓝色二号》绘画的空间组成图(图13-1),并对其进行两种空间立体化处理:其一,对画面中的有机图形给予进深,转化为内部含有空间的体块,获得第一种抽象空间关系图(图13-2、图13-3);其二,对第一种读解方式进行反读,即仅仅保留绘画的边框,将其中有机图形的部分去掉,获得具有中庭的空间感的第二种抽象空间关系图(图13-4、图13-5)。

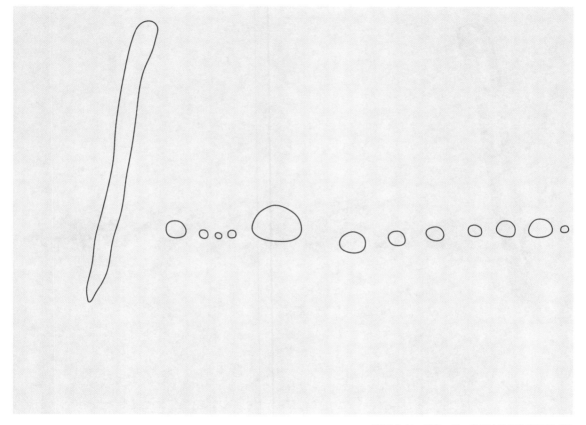

图13-1 从《蓝色二号》绘画中抽取的空间组成图
Fig. 13-1 Spatial composition diagram extracted from *Blue II*

Blue II was completed by Spanish painter Joan Miro in 1961. Miro draws a row of irregular black dots on the blue background, and a vertical red stroke. We read this piece of painting from the perspective of the plane, get the dots-lines relation, and form the spatial composition diagram of *Blue II* (Fig. 13-1). Then we make it three-dimensional in two approaches. One, we add depth to organic graphics, turn them into blocks containing interior spaces, therefore obtain first type of abstract spatial diagrams (Fig. 13-2, Fig. 13-3). Two, we do the opposite, i.e. just keep the painting's frame and remove all the organic graphics, therefore obtain second type of atrium-like abstract spatial diagrams (Fig. 13-4, Fig. 13-5).

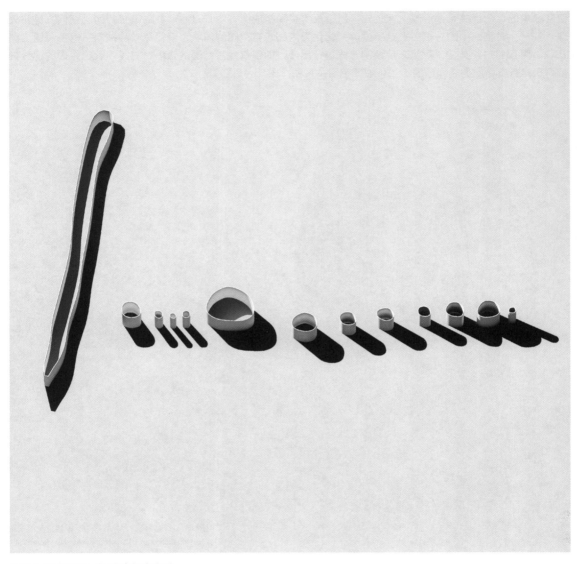

图13-2 《蓝色二号》绘画的空间关系图1
Fig. 13-2 *Blue II*'s spatial diagram 1

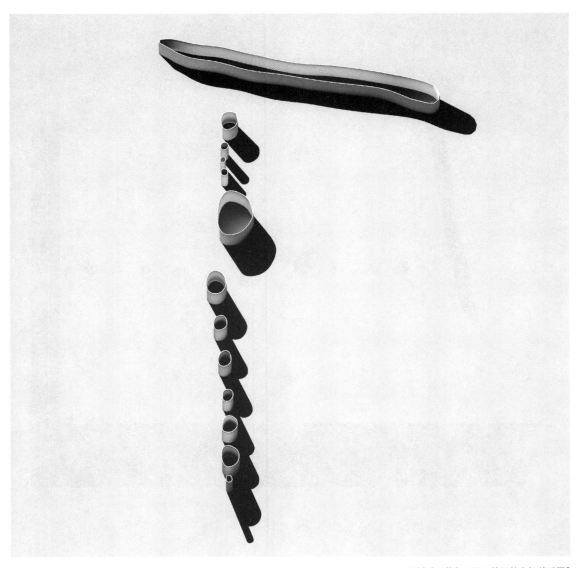

图13-3 《蓝色二号》绘画的空间关系图2
Fig. 13-3 *Blue II*'s spatial diagram 2

图13-4 《蓝色二号》绘画的空间关系图3
Fig. 13-4 *Blue II*'s spatial diagram 3

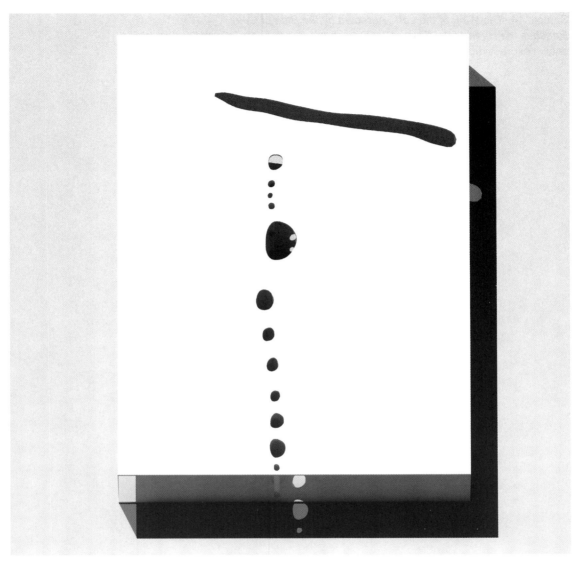

图13-5 《蓝色二号》绘画的空间关系图4
Fig. 13-5 *Blue II*'s spatial diagram 4

实例14《被蓝色光环围绕着的云雀的一只翅膀，
正伸向睡在草地上的罂粟的心脏》绘画的空间性表达
Sample 14 *The Wing of the Lark, Aureoled by the Blue of Gold,
Reaches the Heart of the Poppy Sleeping on the Grass Adorned with Diamonds*'
spatial expression

《被蓝色光环围绕着的云雀的一只翅膀，正伸向睡在草地上的罂粟的心脏》绘画为西班牙画家杰昂·米罗于1967年所绘。画面由一道黑色的笔触在偏向上方的位置划分为两个区域，分别于上、下两个区
The Wing of the Lark, Aureoled by the Blue of Gold, Reaches the Heart of the Poppy Sleeping on the Grass Adorned with Diamonds was completed by Spanish painter Joan Miro in 1967. The picture is divided into two by a black stroke

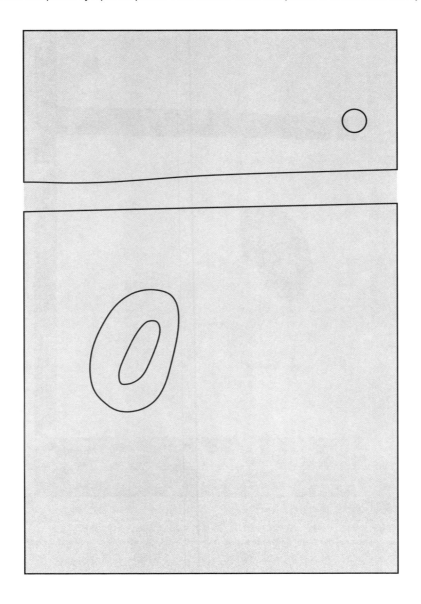

图14-1 从《被蓝色光环围绕着的云雀的一只翅膀，正伸向睡在草地上的罂粟的心脏》绘画中抽取的空间组成图
Fig. 14-1 Spatial composition diagram extracted from *The Wing of the Lark, Aureoled by the Blue of Gold, Reaches the Heart of the Poppy Sleeping on the Grass Adorned with Diamonds*

图14-2《被蓝色光环围绕着的云雀的一只翅膀,正伸向睡在草地上的罂粟的心脏》绘画的空间关系图1
Fig. 14-2 The Wing of the Lark, Aureoled by the Blue of Gold, Reaches the Heart of the Poppy Sleeping on the Grass Adorned with Diamonds' spatial diagram 1

域中绘制了一个较小的圆和一个较大的具有动势的椭圆。我们从平面的视角读解这幅绘画,获取空间中的点、线关系,形成空间组成图(图14-1),进而再对其进行空间立体化处理(给予进深),获得抽象空间关系图(图14-2、图14-3)。

图14-3 《被蓝色光环围绕着的云雀的一只翅膀，正伸向睡在草地上的罂粟的心脏》绘画的空间关系图2
Fig. 14-3 *The Wing of the Lark, Aureoled by the Blue of Gold, Reaches the Heart of the Poppy Sleeping on the Grass Adorned with Diamonds*' spatial diagram 2

in the upper half. There is a relatively smaller circle and bigger dynamic oval respectively. We read this piece of painting from the perspective of the plane, get the dots-lines relation, and form the spatial composition diagram (Fig. 14-1). Then we add depth to make it three-dimensional, therefore obtain the abstract spatial diagrams (Fig. 14-2, Fig. 14-3).

实例15《爵士乐：「礁湖」的插图》
绘画的空间性表达
Sample 15 *Illustration for Jazz: Le Lagon*'s spatial expression

《爵士乐：「礁湖」的插图》绘画为法国画家亨利·马蒂斯于1944年绘制。在这幅绘画中，马蒂斯对具象的对象物进行变形，形成一系列曲线边缘的色块。我们从平面的视角读解这幅绘画，获取空间中的点、线关系，形成《爵士乐：「礁湖」的插图》绘画的空间组成图（图15-1），进而再对其进行空间立体化处理，并划分出虚、实两种不同的空间分割方式，获得《爵士乐：「礁湖」的插图》绘画的抽象空间关系图（图15-2～图15-5）。

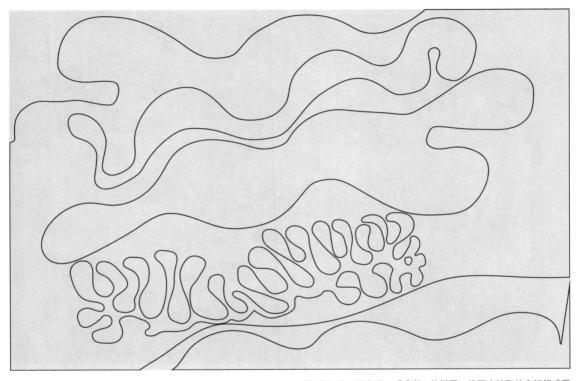

图15-1 从《爵士乐：「礁湖」的插图》绘画中抽取的空间组成图
Fig. 15-1 Spatial composition diagram extracted from *Illustration for Jazz: Le Lagon*

Illustration for Jazz: Le Lagon was completed by French painter Henri Matisse in 1944. In this painting, Matisse transforms the concrete objects to form a series of colour blocks with curved edges. We read this piece of painting from the perspective of the plane, get the dots-lines relation, and form the spatial composition diagram of *Illustration for Jazz: Le Lagon* (Fig. 15-1). Then we make it three-dimensional, divide the work into virtual and actual spatial combinations, therefore obtain *Illustration for Jazz: Le Lagon*'s abstract spatial diagrams (Fig. 15-2 — Fig. 15-5).

图15-2 《爵士乐:「礁湖」的插图》绘画的空间关系图1
Fig. 15-2 *Illustration for Jazz: Le Lagon*'s spatial diagram 1

图15-3《爵士乐:「礁湖」的插图》绘画的空间关系图2
Fig. 15-3 *Illustration for Jazz: Le Lagon*'s spatial diagram 2

图15-4《爵士乐：「礁湖」的插图》绘画的空间关系图3
Fig. 15-4 *Illustration for Jazz: Le Lagon*'s spatial diagram 3

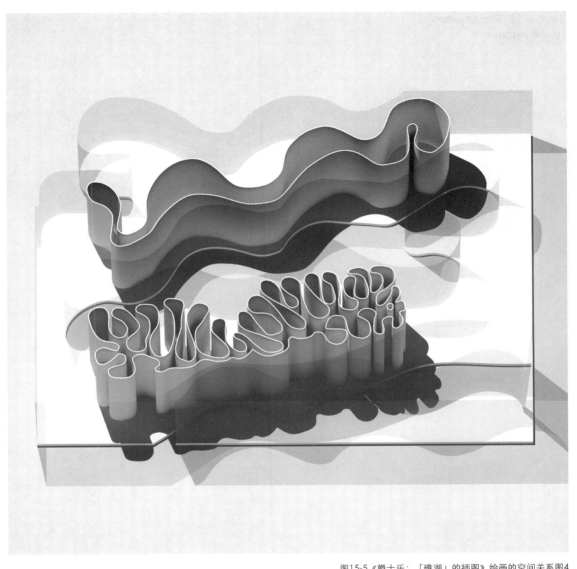

图15-5《爵士乐:「礁湖」的插图》绘画的空间关系图4
Fig. 15-5 *Illustration for Jazz: Le Lagon*'s spatial diagram 4

实例16《爱斯基摩人》
绘画的空间性表达
Sample 16 *The Eskimo*'s
spatial expression

《爱斯基摩人》绘画为法国画家亨利·马蒂斯于1947年绘制。在这幅绘画中，画面在水平方向上被分割成了几个矩形色块，并在每个色块上居中的位置放置了曲线围合而成的图案和一个爱斯基摩人意象变形而成的图形。我们从平面的视角读解这幅绘画，获取空间中的点、线关系，形成《爱斯基摩人》绘画的空间组成图（图16-1），进而再对其进行空间立体化处理，获得《爱斯基摩人》绘画的抽象空间关系图（图16-2、图16-3），并对其中元素进行抽取，获得近似中国园林的空间（图16-4、图16-5）以及空间场景（图16-6、图16-7）。

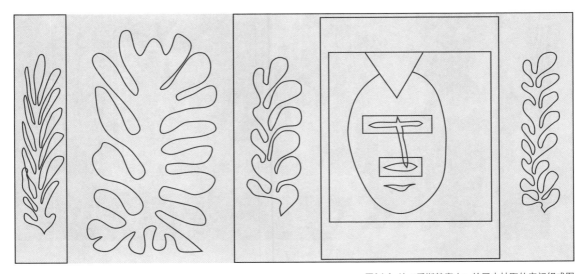

图16-1 从《爱斯基摩人》绘画中抽取的空间组成图
Fig. 16-1 Spatial composition diagram extracted from *The Eskimo*

The Eskimo was completed by French painter Henri Matisse in 1947. In this painting, the picture is horizontally divided into several rectangular colour blocks. There is a pattern enclosed by curves in the centre of every colour block and a figurative image of an Eskimo. We read this piece of painting from the perspective of the plane, get the dots-lines relation, and form the spatial composition diagram of *The Eskimo* (Fig. 16-1). Then we make it three-dimensional, therefore obtain *The Eskimo*'s abstract spatial diagram (Fig. 16-2, Fig 16-3), then get Chinese garden-style space by extracting the elements (Fig. 16-4, Fig. 16-5) and space scenes (Fig. 16-6, Fig. 16-7).

图16-2《爱斯基摩人》绘画的空间关系图1

Fig. 16-2 *The Eskimo*'s spatial diagram 1

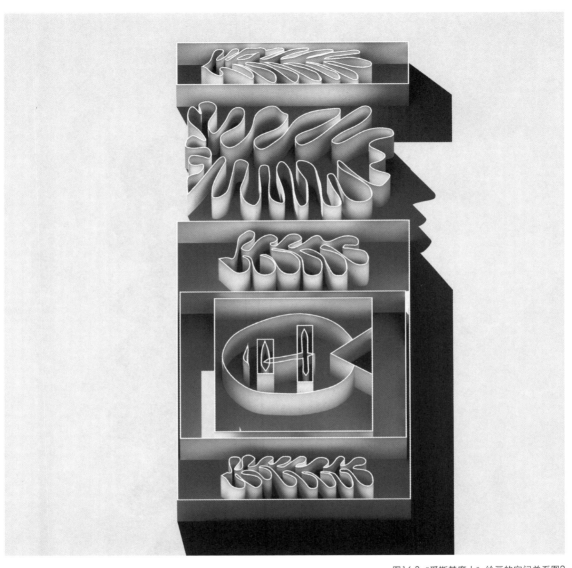

图16-3《爱斯基摩人》绘画的空间关系图2
Fig. 16-3 *The Eskimo*'s spatial diagram 2

图16-4 从《爱斯基摩人》绘画中抽取元素获得的空间关系图1
Fig. 16-4 Spatial diagram of the elements extracted from *The Eskimo* 1

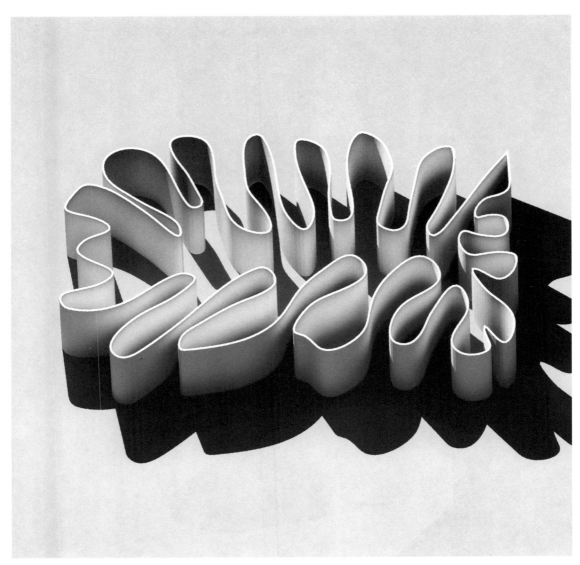

图16-5 从《爱斯基摩人》绘画中抽取元素获得的空间关系图2
Fig. 16-5 Spatial diagram of the elements extracted from *The Eskimo* 2

图16-6 《爱斯基摩人》绘画的空间场景1
Fig. 16-6 Space scenes of *The Eskimo* 1

图16-7 《爱斯基摩人》绘画的空间场景2
Fig. 16-7 Space scenes of *The Eskimo* 2

实例17《头发飘扬的裸像》
绘画的空间性表达
Sample 17 *Nude with Flowing Hair*'s spatial expression

《头发飘扬的裸像》为法国画家亨利·马蒂斯于1952年进行的剪纸创作。画面中表现了一个经过变形的女人体，表达出马蒂斯对人体曲线的理解。这种曲线的轮廓形成了纯粹的面，从而产生了独特的空间体

Nude with Flowing Hair was completed by French painter Henri Matisse in 1952. In this paper-cutting piece, a transformed abstract female body shows Matisse' interpretation of human body curves. The contour of the curves form a pure surface, resulting in a unique spatial experience. We read this piece of painting from the perspective of

图17-1 从《头发飘扬的裸像》绘画中抽取的空间组成图
Fig. 17-1 Spatial composition diagram extracted from *Nude with Flowing Hair*

图17-2《头发飘扬的裸像》绘画的空间关系图1

Fig. 17-2 *Nude with Flowing Hair*'s spatial diagram 1

验。我们从平面的视角读解这幅绘画，获取空间中的点、线关系，形成《头发飘扬的裸像》绘画的空间组成图（图17-1），并对其进行两种空间立体化处理：其一，对画面中的有机图形给予进深，获得第一种抽象空间关系图（图17-2、图17-3）及其场景图（图17-4、图17-5）；其二，对第一种读解方式进行反读，获得具有中庭的空间感的第二种抽象空间关系图（图17-6、图17-7）。

图17-3 《头发飘扬的裸像》绘画的空间关系图2
Fig. 17-3 *Nude with Flowing Hair*'s spatial diagram 2

the plane, get the dots-lines relation, and form the spatial composition diagram of *Nude with Flowing Hair* (Fig. 17-1). Then we make it three-dimensional in two approaches. One, we add depth to organic graphics, turn them into blocks containing interior spaces, therefore obtain first type of abstract spatial diagrams (Fig. 17-2, Fig. 17-3) and space scenes (Fig. 17-4, Fig. 17-5). Two, we do the opposite, i.e. just keep the painting's frame and remove all the organic graphics, therefore obtain second type of atrium-like abstract spatial diagrams (Fig. 17-6, Fig. 17-7).

图17-4《头发飘扬的裸像》绘画的空间场景1
Fig. 17-4 Space scenes of *Nude with Flowing Hair* 1

图17-5《头发飘扬的裸像》绘画的空间场景2
Fig. 17-5 Space scenes of *Nude with Flowing Hair* 2

图17-6 《头发飘扬的裸像》绘画的空间关系图3
Fig. 17-6 *Nude with Flowing Hair*'s spatial diagram 3

图17-7《头发飘扬的裸像》绘画的空间关系图4
Fig. 17-7 Nude with Flowing Hair's spatial diagram 4

实例18《哈马马特的主题》
绘画的空间性表达
Sample 18 *Motif from Hamamet*'s spatial expression

《哈马马特的主题》绘画为瑞士画家保罗·克利于1914年绘制的。在这幅绘画中，画面被分割为一些格子，呈现了一组水彩和以立体主义为基础的半抽象色彩图案的形态，描绘了一种城市、建筑、景观的综合 *Motif from Hamamet* was completed by Swiss painter Paul Klee in 1914. In this piece of painting, the picture is divided into some grids to present a set of semi-abstract watercolour patterns based on Cubism style, depicting a comprehensive image of urban, architecture, and landscape. We read this piece of painting from the perspective of the

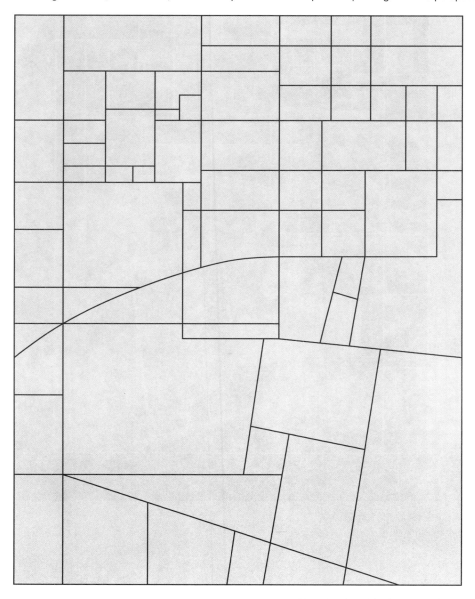

图18-1 从《哈马马特的主题》绘画中抽取的空间组成图

Fig. 18-1 Spatial composition diagram extracted from *Motif from Hamamet*

图18-2 《哈马马特的主题》绘画的空间关系图1

Fig. 18-2 *Motif from Hamamet*'s spatial diagram 1

合面貌。我们从平面的视角读解这幅绘画，获取空间中的点、线关系，形成《哈马马特的主题》绘画的空间组成图（图18-1），进而再对其进行空间立体化处理，获得《哈马马特的主题》绘画的抽象空间关系图（图18-2、图18-3），并对其中元素进行抽取，获得另一种空间读解（图18-4、图18-5）。

图18-3 《哈马马特的主题》绘画的空间关系图2
Fig. 18-3 *Motif from Hamamet*'s spatial diagram 2

plane, get the dots-lines relation, and form the spatial composition diagram of *Motif from Hamamet* (Fig. 18-1). Then we make it three-dimensional, therefore obtain *Motif from Hamamet*'s abstract spatial diagrams (Fig. 18-2, Fig. 18-3), then get another spatial diagrams by extracting the elements (Fig. 18-4, Fig. 18-5).

图18-4 从《哈马马特的主题》绘画中抽取元素获得的空间关系图1
Fig. 18-4 Spatial diagram of the elements extracted from *Motif from Hamamet* 1

图18-5 从《哈马马特的主题》绘画中抽取元素获得的空间关系图2
Fig. 18-5 Spatial diagram of the elements extracted from *Motif from Hamamet* 2

实例19《富饶国土上的纪念碑》
绘画的空间性表达
Sample 19 *Monument in Fertile Country*'s spatial expression

《富饶国土上的纪念碑》绘画为瑞士画家保罗·克利于1929年绘制的。在这幅绘画中，克利用横向线条对画面进行划分，之后用竖向的线条再次切分，通过不同颜色对画面进行分层，这样的操作在画中产生

Monument in Fertile Country was completed by Swiss painter Paul Klee in 1929. In this piece of painting, Klee uses horizontal lines to divide the picture, then split again using vertical lines, and stratify with colors. Such approach

图19-1 从《富饶国土上的纪念碑》绘画中抽取的空间组成图

Fig. 19-1 Spatial composition diagram extracted from Monument in Fertile Country

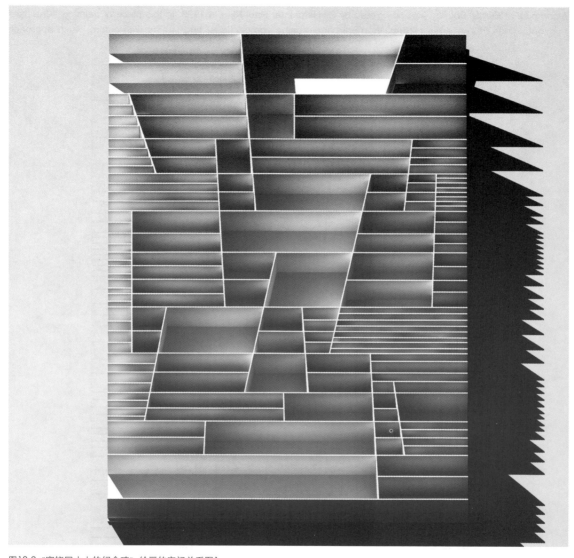

图19-2 《富饶国土上的纪念碑》绘画的空间关系图1

Fig. 19-2 *Monument in Fertile Country*'s spatial diagram 1

一种类似乐谱中的五线谱和各小节的切分的音乐感。我们从平面的视角读解这幅绘画，获取空间中的点、线关系，形成《富饶国土上的纪念碑》绘画的空间组成图（图19-1），进而再对其进行空间立体化处理，获得《富饶国土上的纪念碑》绘画的抽象空间关系图（图19-2～图19-4）。

图19-3 《富饶国土上的纪念碑》绘画的空间关系图2
Fig. 19-3 *Monument in Fertile Country*'s spatial diagram 2

produces a music score style and micro-rhythm. We read this piece of painting from the perspective of the plane, get the dots-lines relation, and form the spatial composition diagram of *Monument in Fertile Country* (Fig. 19-1). Then we make it three-dimensional, therefore obtain *Monument in Fertile Country*'s abstract spatial diagrams (Fig. 19-2 – Fig.19-4).

图19-4《富饶国土上的纪念碑》绘画的空间关系图3
Fig. 19-4 *Monument in Fertile Country*'s spatial diagram 3

实例20 《傍晚的火》
绘画的空间性表达
Sample 20 *Fire in Evening*'s spatial expression

《傍晚的火》绘画为瑞士画家保罗·克利于1929年绘制的。在这幅绘画中，克利用横向线条对画面进行划分，之后用竖向的线条再次切分，通过不同颜色对画面进行分层，这样的操作在画中产生一

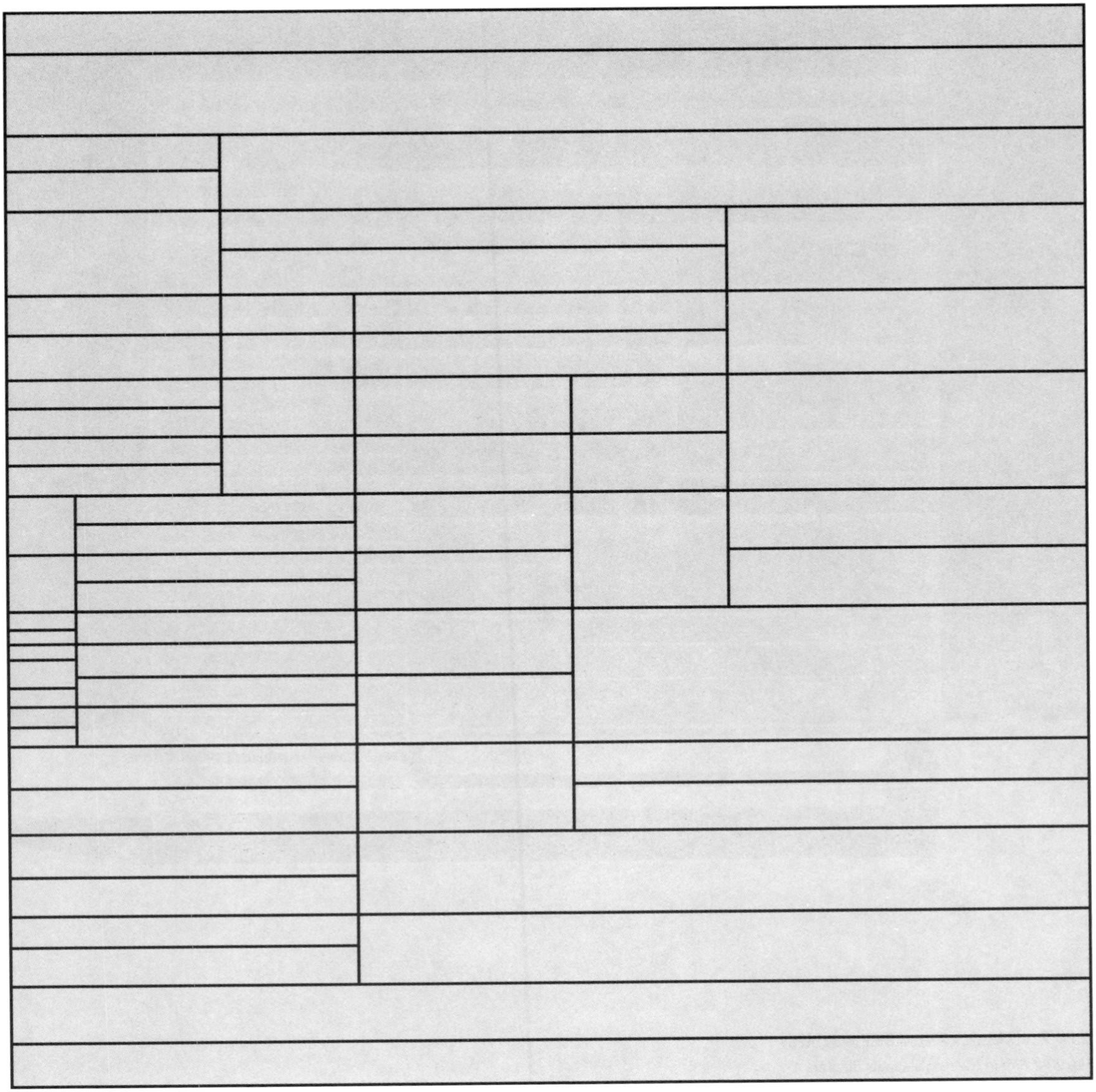

图20-1 从《傍晚的火》绘画中抽取的空间组成图
Fig. 20-1 Spatial composition diagram extracted from *Fire in Evening*

Fire in Evening was completed by Swiss painter Paul Klee in 1929. In this piece of painting, Klee uses horizontal lines to divide the picture, then split again using vertical lines, and stratify with colors. Such approach produces a music

图20-2 《傍晚的火》绘画的空间关系图1
Fig. 20-2 *Fire in Evening*'s spatial diagram 1

种类似乐谱中的五线谱和各小节的切分的音乐感。我们从立面和剖面的视角读解这幅绘画，获取空间中的点、线关系，形成《傍晚的火》绘画的空间组成图（图20-1），进而再对其进行空间立体化处理（给予进深），获得《傍晚的火》绘画的抽象空间关系图（图20-2、图20-3）。

图20-3 《傍晚的火》绘画的空间关系图2
Fig. 20-3 *Fire in Evening*'s spatial diagram 2

score style and micro-rhythm. We read this piece of painting from the perspectives of elevation and section, get the dots-lines relation, and form the spatial composition diagram of *Fire in Evening* (Fig. 20-1). We add depth to make it three-dimensional, therefore obtain *Fire in Evening*'s abstract spatial diagrams (Fig. 20-2, Fig. 20-3).

实例21《岩石间的小镇》
绘画的空间性表达
Sample 21 *Small Town among the Rocks*' spatial expression

《岩石间的小镇》绘画为瑞士画家保罗·克利于1932年绘制的。折线将画面切碎,位于画面中上似乎存在一条水平的直线将画面分割为上、下两部分、上部的几何碎片形成类似立面的组合,下部的几何碎片形成类似平面的组合。我们从立面和剖面的视角读解其上部,从平面的视角读解其下部,获取空间中的点、线关系,形成空间组成图(图21-1),并对其进行空间立体化处理,对两部分给予进深,使其在相互垂直的方向上叠加,获得《岩石间的小镇》绘画的抽象空间关系图(图21-2~图21-4)。

图21-1 从《岩石间的小镇》绘画中抽取的空间组成图
Fig. 21-1 Spatial composition diagram extracted from *Small Town among the Rocks*

Small Town among the Rocks was completed by Swiss painter Paul Klee in 1932. The polyline chops the picture into pieces. There appears to be a horizontal line dividing the picture into upper and lower portions. Geometrical fragments in the upper part form a facade-like combination, and geometrical fragments in lower part form a plane-like combination. We read the upper image of painting from the perspectives of elevation and section, the lower from the plane, get the dots-lines relation, and form the spatial composition diagram (Fig. 21-1). Then we add depth to both parts to make it three-dimensional and orthogonally superposed on one another, therefore obtain *Small Town among the Rocks*' abstract spatial diagrams (Fig. 21-2 – Fig. 21-4).

图21-2《岩石间的小镇》绘画的空间关系图1
Fig. 21-2 *Small Town among the Rocks*' spatial diagram 1

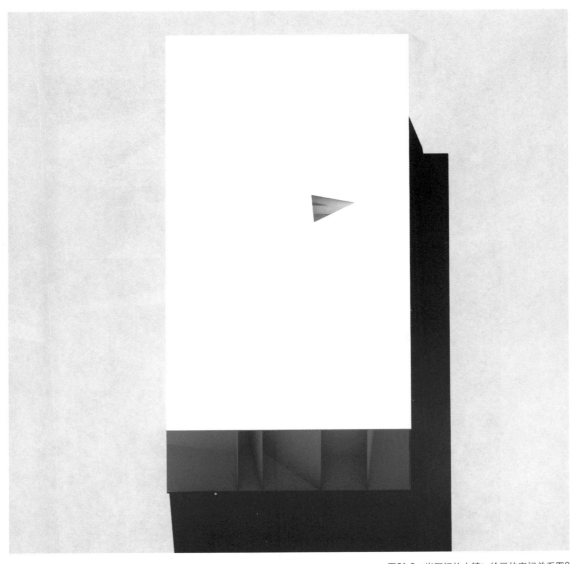

图21-3 《岩石间的小镇》绘画的空间关系图2
Fig. 21-3 *Small Town among the Rocks*' spatial diagram 2

图21-4《岩石间的小镇》绘画的空间关系图3
Fig. 21-4 *Small Town among the Rocks*' spatial diagram 3

实例22《蓝色的夜》
绘画的空间性表达
Sample 22 *Blue Night*'s spatial expression

《蓝色的夜》绘画为瑞士画家保罗·克利于1937年绘制的。画面中通过一些自由放置的直线、曲线以及不同颜色的色块从两个层次上对空间进行切分。我们从平面的视角读解这幅绘画，获取空间中的点、线关系，形成《蓝色的夜》绘画的空间组成图（图22-1），进而再对其进行空间立体化处理，将结构线条给予进深生成墙体，并根据色块对形成的空间进行再次分割，形成具有近似房屋、街道、中庭等意象的空间，获得《蓝色的夜》绘画的抽象空间关系图（图22-2、图22-3）。

图22-1 从《蓝色的夜》绘画中抽取的空间组成图
Fig. 22-1 Spatial composition diagram extracted from *Blue Night*

Blue Night was completed by Swiss painter Paul Klee in 1937. Through a number of free-standing lines, curves and different coloured blocks divide the picture's spaces on two levels. We read this piece of painting from the perspective of the plane, get the dots-lines relation, and form the spatial composition diagram of *Blue Night* (Fig. 22-1). Then we add depth to the structural lines to make it walls, spilt the coloured block spaces again, form figurative spaces like buildings, streets, and atriums, and therefore obtain abstract spatial diagrams (Fig. 22-2, Fig. 22-3).

图22-2 《蓝色的夜》绘画的空间关系图1
Fig. 22-2 *Blue Night*'s spatial diagram 1

图22-3 《蓝色的夜》绘画的空间关系图2
Fig. 22-3 *Blue Night*'s spatial diagram 2

实例23《和谐的战场》
绘画的空间性表达
Sample 23 *Harmonized Battle*'s spatial expression

《和谐的战场》绘画为瑞士画家保罗·克利于1937年绘制的。画面中通过一些自由放置的直线以及不同颜色的色块从两个层次上对空间进行切分。我们从平面的视角读解这幅绘画,获取空间中的点、线关系,形成《和谐的战场》绘画的空间组成图(图23-1),进而再对其进行空间立体化处理,将结构线条给予进深生成墙体,并根据色块对形成的空间进行再次分割,形成具有近似房屋、街道、中庭等意象的空间,获得《和谐的战场》绘画的抽象空间关系图(图23-2、图23-3)。

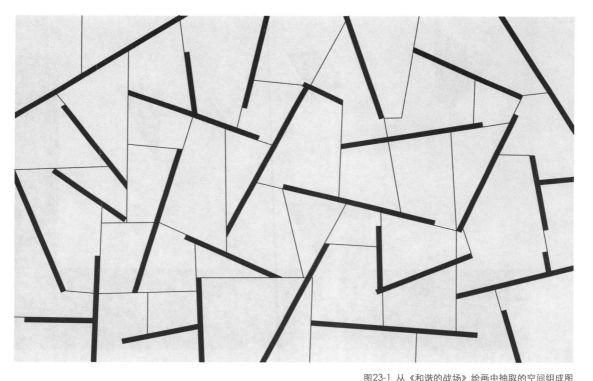

图23-1 从《和谐的战场》绘画中抽取的空间组成图
Fig. 23-1 Spatial composition diagram extracted from *Harmonized Battle*

Harmonized Battle was completed by Swiss painter Paul Klee in 1937. Through a number of free-standing lines and different coloured blocks, the picture's spaces are divided on two levels. We read this piece of painting from the perspective of the plane, get the dots-lines relation, and form the spatial composition diagram of *Harmonized Battle* (Fig. 23-1). Then we add depth to the structural lines to make it walls, spilt the coloured block spaces again, form figurative spaces like buildings, streets, and atriums, and therefore obtain *Harmonized Battle*'s abstract spatial diagrams (Fig. 23-2, Fig. 23-3).

图23-2 《和谐的战场》绘画的空间关系图1
Fig. 23-2 *Harmonized Battle*'s spatial diagram 1

图23-3 《和谐的战场》绘画的空间关系图2
Fig. 23-3 *Harmonized Battle*'s spatial diagram 2

实例24《构成第八号》
绘画的空间性表达
Sample 24 *Composition VIII*'s spatial expression

《构成第八号》绘画为俄罗斯画家瓦西里·康定斯基于1923年绘制的。这幅绘画由三角形、圆形等几何形式以及不稳定的斜线等元素构成,各种元素之间相互交错叠加,于结构上具有强烈的动势。我们从平面的视角读解这幅绘画,获取空间中的点、线关系,形成《构成第八号》绘画的空间组成图(图24-1),进而再对其进行空间立体化处理,获得《构成第八号》绘画的抽象空间关系图(图24-2)及其空间场景图(图24-3、图24-4)。

图24-1 从《构成第八号》绘画中抽取的空间组成图
Fig. 24-1 Spatial composition diagram extracted from *Composition VIII*

Composition VIII was completed by Russian painter Wassily Kandinsky in 1923. This piece of painting is composed of triangles, circles and other geometric forms, as well as unstable elements like slashes. All elements juxtapose and form a strong structural momentum. We read this piece of painting from the perspective of the plane, get the dots-lines relation, and form the spatial composition diagram of *Composition VIII* (Fig. 24-1). Then we add depth to make it three-dimensional, therefore obtain *Composition VIII*'s abstract spatial diagrams (Fig. 24-2) and space scenes (Fig. 24-3, Fig. 24-4).

图24-2《构成第八号》绘画的空间关系图
Fig. 24-2 *Composition VIII*'s spatial diagram

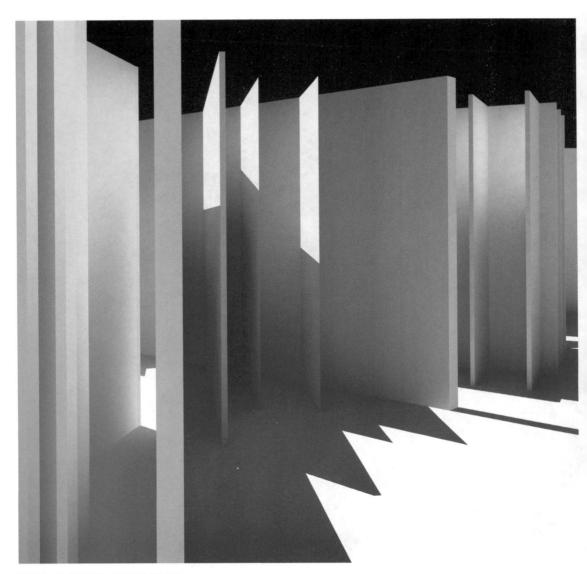

图24-3 《构成第八号》绘画的空间场景1
Fig. 24-3 Space scenes of *Composition VIII* 1

图24-4 《构成第八号》绘画的空间场景2
Fig. 24-4 Space scenes of *Composition VIII* 2

实例25《黄·红·蓝》
绘画的空间性表达
Sample 25 *Yellow-Red-Blue*'s spatial expression

《黄·红·蓝》绘画为俄罗斯画家瓦西里·康定斯基于1925年绘制的。这幅绘画由三角形、圆形等几何形式以及自由的曲线等有机图形构成,各种元素之间相互交错叠加,使结构上具有强烈的动势。我们从平面的视角读解这幅绘画,获取空间中的点、线关系,形成《黄·红·蓝》绘画的空间组成图(图25-1),进而再对其进行空间立体化处理,获得《黄·红·蓝》绘画的抽象空间关系图(图25-2)及其空间场景图(图25-3、图25-4)。

图25-1 从《黄·红·蓝》绘画中抽取的空间组成图

Fig. 25-1 Spatial composition diagram extracted from *Yellow-Red-Blue*

Yellow-Red-Blue was completed by Russian painter Wassily Kandinsky in 1925. This piece of painting is composed of triangles, circles and other geometric forms, as well as unstable elements like slashes. All elements juxtapose and form a strong structural momentum. We read this piece of painting from the perspective of the plane, get the dots-lines relation, and form the spatial composition diagram of *Yellow-Red-Blue* (Fig. 25-1). Then we add depth to make it three-dimensional, therefore obtain *Yellow-Red-Blue*'s abstract spatial diagrams (Fig. 25-2) and space scenes (Fig. 25-3, Fig. 25-4).

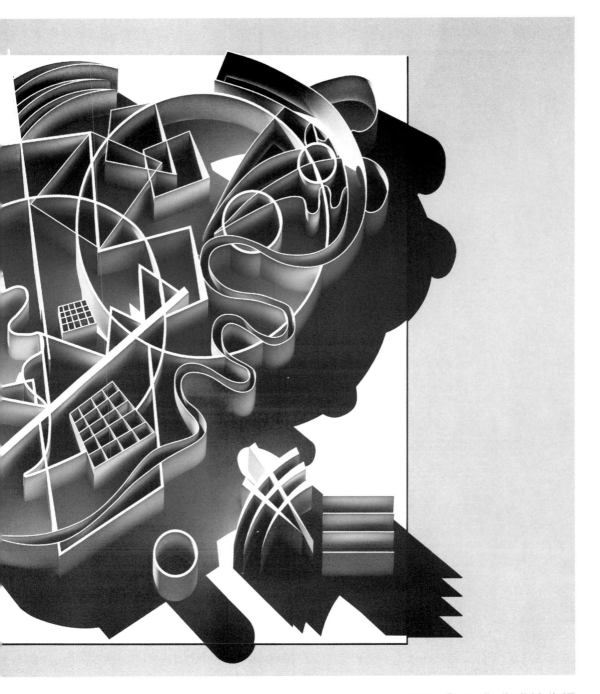

图25-2 《黄·红·蓝》绘画的空间关系图
Fig. 25-2 *Yellow-Red-Blue*'s spatial diagram

图25-3 《黄·红·蓝》绘画的空间场景1
Fig. 25-3 Space scenes of *Yellow-Red-Blue* 1

图25-4 《黄·红·蓝》绘画的空间场景2
Fig. 25-4 Space scenes of *Yellow-Red-Blue* 2

实例26《粉红色的音调》
绘画的空间性表达
Sample 26 *Accent in Pink*'s spatial expression

《粉红色的音调》绘画为俄罗斯画家瓦西里·康定斯基于1926年所绘。画面被一个边线为曲线且中心别切去一个矩形的四边形分割成外部、中部、内部三个部分,并在另一个层次上布置了散落的、相互叠加

Accent in Pink was completed by Russian painter Wassily Kandinsky in 1926. The picture is divided into three parts - external, middle and internal - by a rectangle with curve-shaped outline and cropped by a rectangular. Several

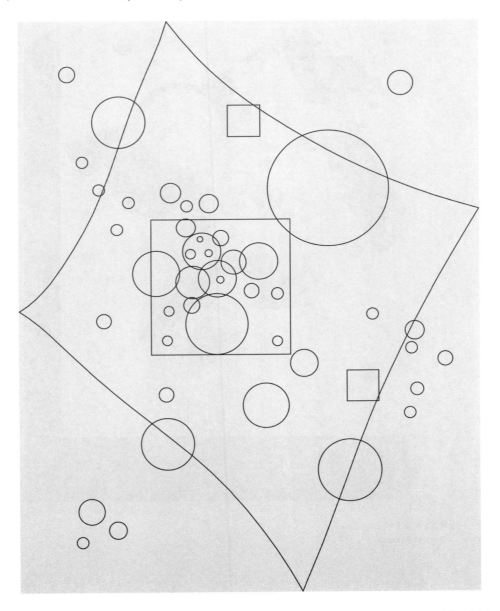

图26-1 从《粉红色的音调》绘画中抽取的空间组成图

Fig. 26-1 Spatial composition diagram extracted from *Accent in Pink*

图26-2 《粉红色的音调》绘画的空间关系图1
Fig. 26-2 *Accent in Pink*'s spatial diagram 1

交错的圆形个几个矩形色块。我们从平面的视角读解这幅绘画，获取空间中的点、线关系，形成《粉红色的音调》绘画的空间组成图（图26-1），进而再对其进行空间立体化处理，获得《粉红色的音调》绘画的抽象空间关系图（图26-2、图26-3）。

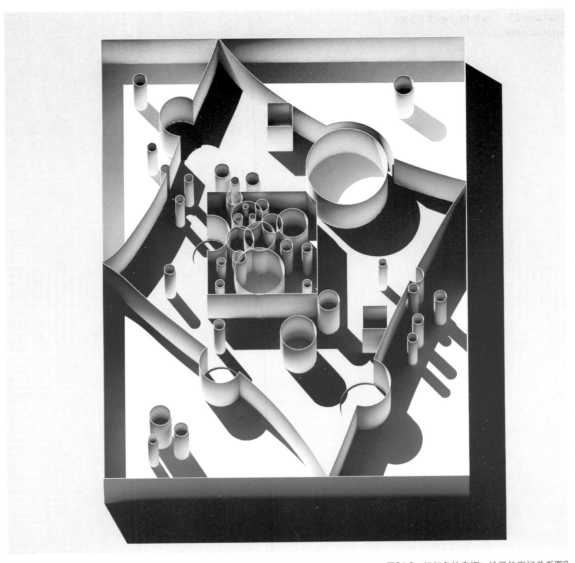

图26-3 《粉红色的音调》绘画的空间关系图2
Fig. 26-3 *Accent in Pink*'s spatial diagram 2

scattered rectangle coloured blocks are placed on another layer of superimposed circles. We read this piece of painting from the perspective of the plane, get the dots-lines relation, and form the spatial composition diagram of *Accent in Pink* (Fig. 26-1). Then we add depth to make it three-dimensional, therefore obtain *Accent in Pink*'s abstract spatial diagrams (Fig. 26-2, Fig. 26-3).

实例27《十三个矩形》
绘画的空间性表达
Sample 27 *Thirteen Rectangles'* spatial expression

《十三个矩形》绘画为俄罗斯画家瓦西里·康定斯基于1930年绘制的。在这幅绘画中,康定斯基绘制了13个相互连续、交错、叠加的矩形色块。我们从平面的视角对这幅绘画进行读解,获取空间中的点、线

Thirteen Rectangles was completed by Russian painter Wassily Kandinsky in 1930. In this piece of painting, Kandinsky draws 13 continuous superimposed rectangle colour blocks. We read this piece of painting from the perspective of the

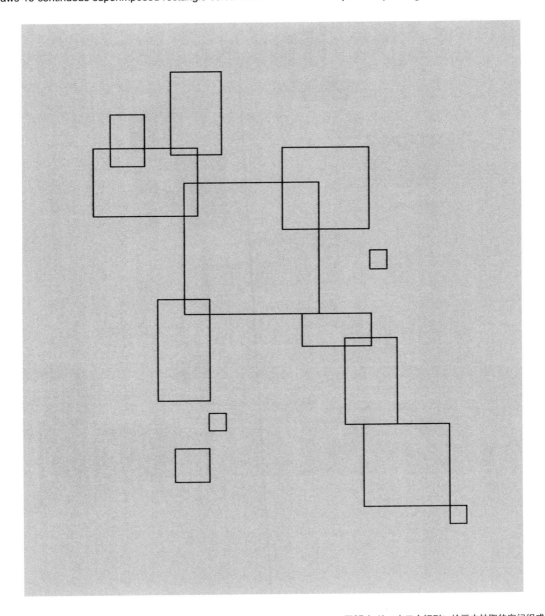

图27-1 从《十三个矩形》绘画中抽取的空间组成图

Fig. 27-1 Spatial composition diagram extracted from Thirteen Rectangles

图27-2 《十三个矩形》绘画的空间关系图1
Fig. 27-2 *Thirteen Rectangles*' spatial diagram 1

关系，形成《十三个矩形》绘画的空间组成图（图27-1），在此基础上，进而对其进行空间立体化处理，获得《十三个矩形》绘画的抽象空间关系图（图27-2、图27-3）。

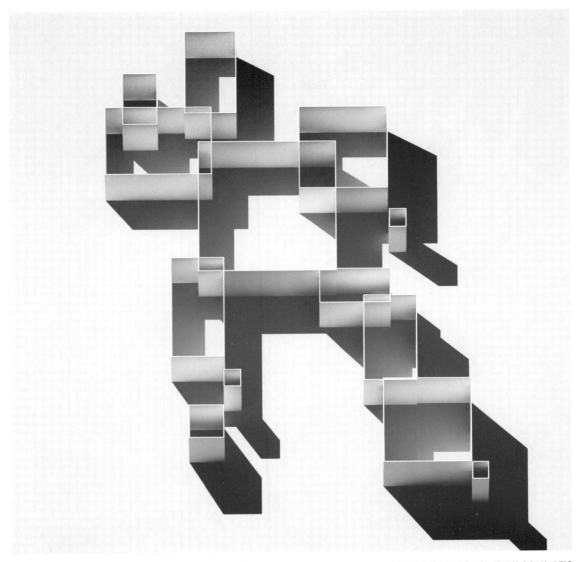

图27-3 《十三个矩形》绘画的空间关系图2
Fig. 27-3 *Thirteen Rectangles*' spatial diagram 2

plane, get the dots-lines relation, and form the spatial composition diagram of Thirteen Rectangles (Fig. 27-1). Then we add depth to make it three-dimensional, therefore obtain *Thirteen Rectangles*' abstract spatial diagrams (Fig. 27-2, Fig. 27-3).

实例28《难以忍受的张力》
绘画的空间性表达
Sample 28 *Hard Tension*'s
spatial expression

《难以忍受的张力》绘画为俄罗斯画家瓦西里·康定斯基于1931年所绘。画面由矩形、圆形、梯形以及具有一定宽度的直线组成,并沿这些几何形体的外围围合了一个边缘为曲线的深色色块。我们从平面的

Hard Tension was completed by Russian painter Wassily Kandinsky in 1931. The painting is composed of rectangle, circle, trapezoid, and lines with certain thickness. A curve-edged dark-color block encloses these geometrical forms.

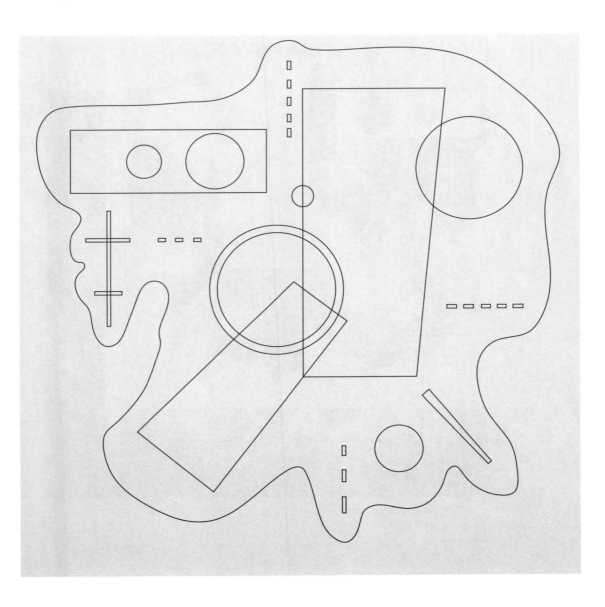

图28-1 从《难以忍受的张力》绘画中抽取的空间组成图
Fig. 28-1 Spatial composition diagram extracted from *Hard Tension*

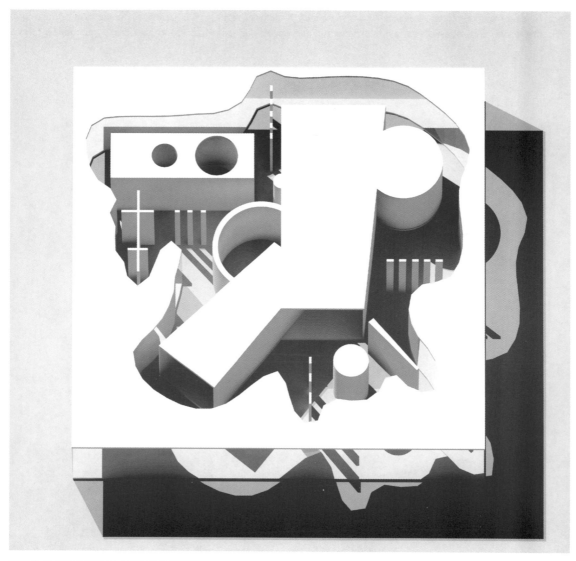

图28-2 《难以忍受的张力》绘画的空间关系图1
Fig. 28-2 *Hard Tension*'s spatial diagram 1

视角读解这幅绘画，获取空间中的点、线关系，形成《难以忍受的张力》绘画的空间组成图（图28-1），进而再对其进行空间立体化处理，获得《难以忍受的张力》绘画的抽象空间关系图（图28-2、图28-3）。

图28-3 《难以忍受的张力》绘画的空间关系图2
Fig. 28-3 *Hard Tension*'s spatial diagram 2

We read this piece of painting from the perspective of the plane, get the dots-lines relation, and form the spatial composition diagram of *Hard Tension* (Fig. 28-1). Then we add depth to make it three-dimensional, therefore obtain *Hard Tension*'s abstract spatial diagrams (Fig. 28-2, Fig. 28-3).

实例29《无缘无故的上升》
绘画的空间性表达
Sample 29 *Gratuitous Ascent*'s spatial expression

《无缘无故的上升》绘画为俄罗斯画家瓦西里·康定斯基于1934年绘制的。画面中以正交的水平、竖直直线为轴线，分布了圆形、矩形等几何形体以及细胞式的有机形态。我们从平面的视角读解这幅绘

Gratuitous Ascent was completed by Russian painter Wassily Kandinsky in 1934. The picture uses orthogonal axis to distribute circle, rectangle and other geometric forms, as well as cell-type organic forms. We read this piece of painting

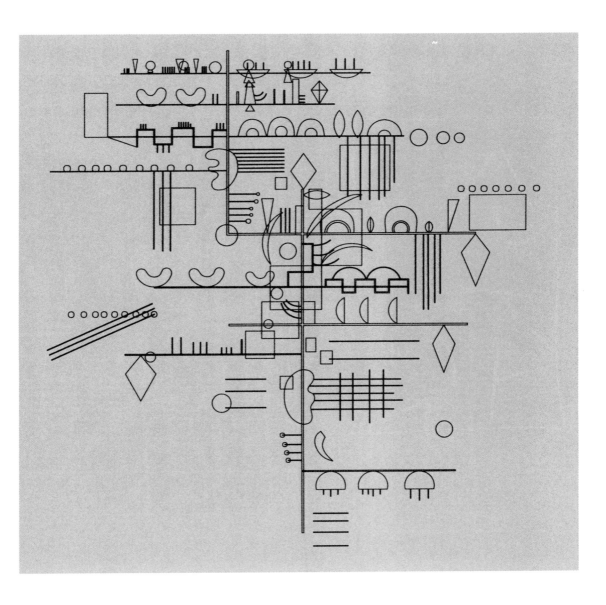

图29-1 从《无缘无故的上升》绘画中抽取的空间组成图
Fig. 29-1 Spatial composition diagram extracted from *Gratuitous Ascent*

图29-2 《无缘无故的上升》绘画的空间关系图1
Fig. 29-2 *Gratuitous Ascent*'s spatial diagram 1

画,获取空间中的点、线关系,形成《无缘无故的上升》绘画的空间组成图(图29-1),进而再对其进行空间立体化处理,获得一种近似于城市尺度上的《无缘无故的上升》绘画的抽象空间关系图(图29-2、图29-3)及其空间场景图(图29-4、图29-5)。

图29-3 《无缘无故的上升》绘画的空间关系图2
Fig. 29-3 *Gratuitous Ascent*'s spatial diagram 2

from the perspective of the plane, get the dots-lines relation, and form the spatial composition diagram of *Gratuitous Ascent* (Fig. 29-1). Then we add depth to make it three-dimensional, therefore obtain *Gratuitous Ascent*'s abstract spatial diagrams at urban scale (Fig. 29-2, Fig. 29-3) and space scenes (Fig. 29-4, Fig. 29-5).

图29-4 《无缘无故的上升》绘画的空间场景1
Fig. 29-4 Space scenes of *Gratuitous Ascent* 1

图29-5 《无缘无故的上升》绘画的空间场景2
Fig. 29-5 Space scenes of *Gratuitous Ascent* 2

实例30《画了节拍器的静物》
绘画的空间性表达
Sample 30 *Still Life with Metronome*'s spatial expression

《画了节拍器的静物》为法国画家乔治·布拉克于1910年绘制的立体主义绘画。其运用分析立体主义绘画的操作方法，以一组客观存在的静物作为对象物，将其几何学化，并对其从多视点连续观察，将观察

图30-1 从《画了节拍器的静物》绘画中抽取的空间组成图
Fig. 30-1 Spatial composition diagram extracted from *Still Life with Metronome*

Still Life with Metronome was completed by French Cubism painter Georges Braque in 1910. By means of Cubism painting analytics, Georges Braque turns a set of still life objects into geometrical forms, conducts continuous observa-

图30-2 《画了节拍器的静物》绘画的空间俯视图

Fig. 30-2 *Still Life with Metronome*'s bird view

所得的三维空间叠加于二维的绘画平面中。整幅绘画处于一个统一的色调中，物体的体量感以及空间的进深感被削弱。我们从平面的视角读解这幅绘画，获取空间中的点、线关系，形成《画了节拍器的静物》绘画的空间组成图（图30-1），并对其进行空间立体化处理，获得《画了节拍器的静物》绘画的空间整体俯视图（图30-2）以及抽象空间关系图（图30-3、图30-4）。

图30-3 《画了节拍器的静物》绘画的空间关系图1
Fig. 30-3 *Still Life with Metronome*'s spatial diagram 1

tions from various viewpoints, superimposes the three-dimensional spaces on the two-dimensional painting. The whole piece has a unified tone, thus the objects' volume and spatial depth is weakened. We read this piece of painting from the perspective of the plane, get the dots-lines relation, and form the spatial composition diagram of *Still Life with Metronome* (Fig. 30-1). Then we add depth to make it three-dimensional, therefore obtain *Still Life with Metronome*'s bird view (Fig. 30-2) and abstract spatial diagrams (Fig. 30-3, Fig. 30-4).

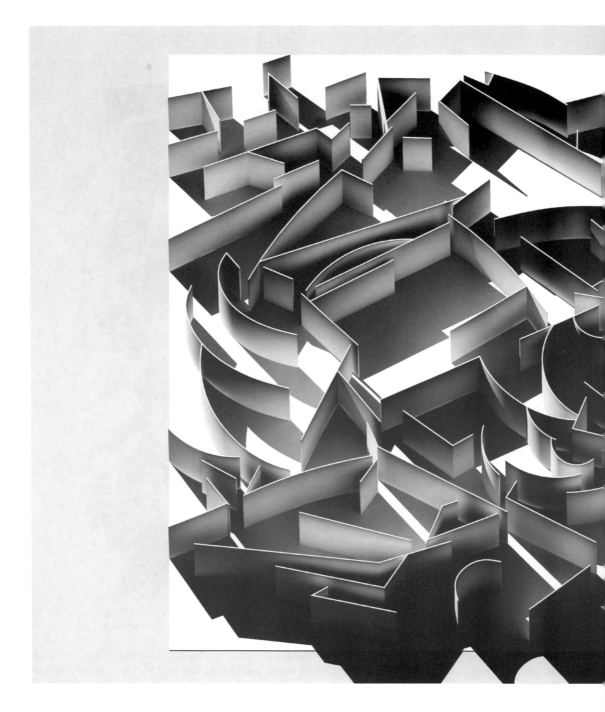

图30-4 《画了节拍器的静物》绘画的空间关系图2
Fig. 30-4 *Still Life with Metronome*'s spatial diagram 2

实例31《小提琴》
绘画的空间性表达
Sample 31 *The Violin*'s spatial expression

《小提琴》为法国画家乔治·布拉克于1911年绘制的立体主义绘画。这幅绘画在应用于《画了节拍器的静物》绘画中的"多视点连续观察""将观察所得的三维空间叠加于二维的绘画平面中"的绘画方法的

图31-1 从《小提琴》绘画中抽取的空间组成图
Fig. 31-1 Spatial composition diagram extracted from *The Violin*

The Violin was completed by French Cubism painter Georges Braque in 1911. Based on the "multi-view continuous observation" and "superimposing three-dimensional spaces on two-dimensional painting" approaches of *Still Life with*

图31-2 《小提琴》绘画的空间俯视图
Fig. 31-2 *The Violin*'s bird view

基础上,突出了对线条的表现。我们从平面的视角读解这幅绘画,获取空间中的点、线关系,形成《小提琴》绘画的空间组成图(图31-1),并对其进行空间立体化处理,获得《小提琴》绘画的空间整体俯视图(图31-2)以及抽象空间关系图(图31-3、图31-4)。

图31-3 《小提琴》绘画的空间关系图1
Fig. 31-3 *The Violin*'s spatial diagram 1

Metronome, this piece is more stressed on the expression of lines. We read this piece of painting from the perspective of the plane, get the dots-lines relation, and form the spatial composition diagram of *The Violin* (Fig. 31-1). Then we add depth to make it three-dimensional, therefore obtain *The Violin*'s bird view (Fig. 31-2) and abstract spatial diagrams (Fig. 31-3, Fig. 31-4).

图31-4《小提琴》绘画的空间关系图2
Fig. 31-4 *The Violin*'s spatial diagram 2

实例32《有玻璃杯和报纸的静物》
绘画的空间性表达
Sample 32 *Still Life with Glass and Newspaper*'s spatial expression

《有玻璃杯和报纸的静物》为法国画家乔治·布拉克于1913年绘制的立体主义绘画。与前文提到的两幅绘画相比，在这幅绘画中，布拉克进行了对椭圆形边缘画面的探索，并且进一步突出了对线条的表现，

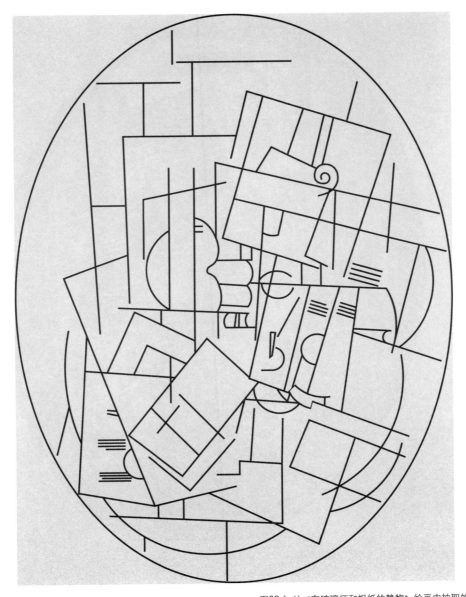

图32-1 从《有玻璃杯和报纸的静物》绘画中抽取的空间组成图
Fig. 32-1 Spatial composition diagram extracted from *Still Life with Glass and Newspaper*

Still Life with Glass and Newspaper was completed by French Cubism painter Georges Braque in 1913. Compared with the previously mentioned two paintings, Georges Braque explores on the oval edge, and further highlights the

图32-2 《有玻璃杯和报纸的静物》绘画的空间俯视图

Fig. 32-2 *Still Life with Glass and Newspaper*'s bird view

同时,画面中还出现了符号性的文字内容。我们从平面的视角读解这幅绘画,获取空间中的点、线关系,形成《有玻璃杯和报纸的静物》绘画的空间组成图(图32-1),进而再对其进行空间立体化处理,获得《有玻璃杯和报纸的静物》绘画的空间整体俯视图(图32-2)以及抽象空间关系图(图32-3、图32-4)。

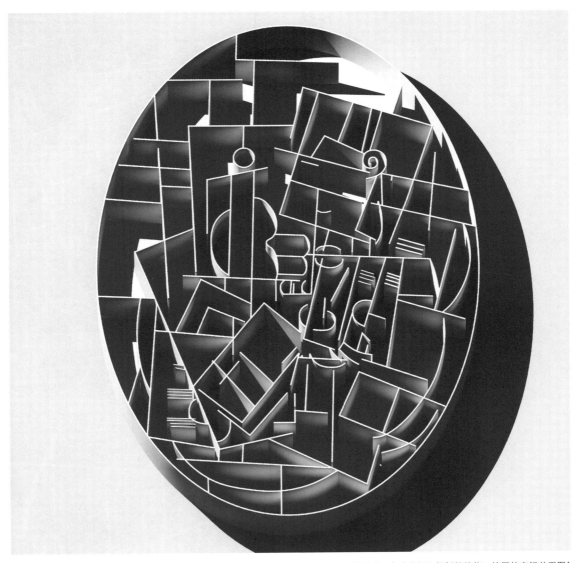

图32-3 《有玻璃杯和报纸的静物》绘画的空间关系图1
Fig. 32-3 *Still Life with Glass and Newspaper*'s spatial diagram 1

expression of lines. At the same time, symbolic texts appear in the picture. We read this piece of painting from the perspective of the plane, get the dots-lines relation, and form the spatial composition diagram of *Still Life with Glass and Newspaper* (Fig. 32-1). Then we add depth to make it three-dimensional, therefore obtain *Still Life with Glass and Newspaper*'s bird view (Fig. 32-2) and abstract spatial diagrams (Fig. 32-3, Fig. 32-4).

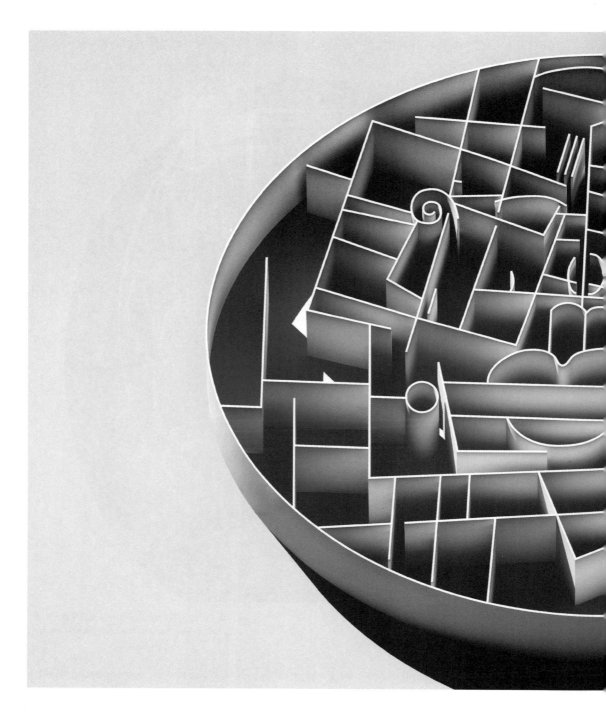

图32-4《有玻璃杯和报纸的静物》绘画的空间关系图2
Fig. 32-4 *Still Life with Glass and Newspaper*'s spatial diagram 2

空间的叠加的示例
Spatial superposition sample

　　在这一章节中，我们重点地对蒙德里安的三幅绘画：《狐步舞A》《有黑色线条的构图II》《有两条线的构图》中的三个空间进行分析，并以叠加的方式对其进行空间性试做。
In this chapter, we will focus on Mondrian's paintings - *Fox Trot A*, *Lozenge Composition with Two Lines*, *Composition II with Black Lines*. We analyse three spaces, and use superimposition approach to conduct experiments.

实例33《狐步舞A》《有黑色线条的构图Ⅱ》《有两条线的构图》
绘画叠加的空间性表达
Sample 33 Juxtaposition spatial expression of *Fox Trot A*, *Composition II with Black Lines*, and *Lozenge Composition with Two Lines*

《狐步舞A》《有黑色线条的构图 II》《有两条线的构图》为荷兰画家彼埃·蒙德里安于1930年和1931年绘制的构成主义绘画。在这几幅绘画中，水平、竖直的几条具有一定宽度的正交结构线条将画面在平面上进行了分割。我们从平面的视角读解这几幅绘画，所获得的空间是简洁的。但这种黑色的线更像建筑图纸中的墙，尽管我们可以将这个墙想成街道而形成空间关系图1（图33-1）的图底关系，但似乎职业性的习惯，这几条黑线在我眼里还是一组如空间关系图2（图33-2）中所示的"厚墙"。我们在空间构成图2基础上对其进行空间立体化处理，分别获得三幅绘画的抽象空间关系图（图34-3），再将抽象出的空间在竖直方向上进行叠加，形成叠加后的抽象空间关系图（图33-4~图33-8）。

图33-1 从《狐步舞A》《有黑色线条的构图 II》《有两条线的构图》绘画中抽取的空间组成图1

Fig. 33-1 Spatial composition diagram extracted from *Fox Trot A*, *Composition II with Black Lines*, *Lozenge Composition with Two Lines* 1

图33-2 从《狐步舞A》《有黑色线条的构图 II》《有两条线的构图》绘画中抽取的空间组成图2

Fig. 33-2 Spatial composition diagram extracted from *Fox Trot A*, *Composition II with Black Lines*, *Lozenge Composition with Two Lines* 2

Fox Trot A, *Composition II with Black Lines*, *Lozenge Composition with Two Lines* were completed by Dutch painter Piet Mondrian in 1930 and 1931 respectively. In these Cubism paintings, orthogonal structural lines with certain thickness divide the picture. We read this piece of painting from the perspective of the plane, and get concise space. Yet these black lines are more like walls in architecture drawings. Although we can think these walls as streets to form spatial diagram 1's (Fig. 33-1) figure-ground relation, out of occupational habits, these black lines seem to me a set of "thick walls" in spatial diagram 2 (Fig. 33-2). On the basis of spatial diagram 2, we make it three-dimensional, therefore obtain abstract spatial diagrams of the three paintings (Fig. 33-3). We then vertically superimpose the extracted spaces to form an abstract spatial diagrams (Fig. 33-4 — Fig. 33-8).

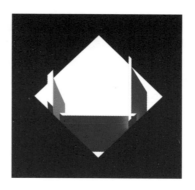
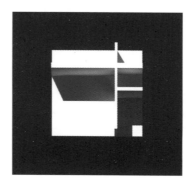

图33-3《狐步舞A》《有黑色线条的构图Ⅱ》《有两条线的构图》绘画的空间关系图1
Fig. 33-3 *Fox Trot A, Composition II with Black Lines, Lozenge Composition with Two Lines*' spatial diagram 1

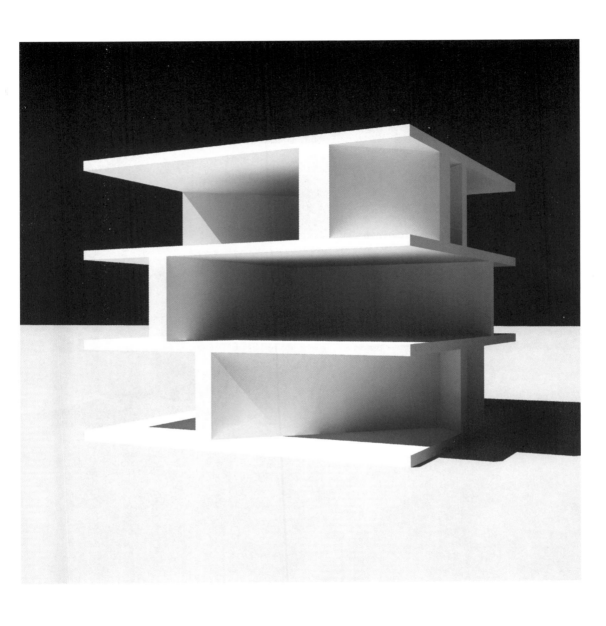

图33-4《狐步舞A》《有黑色线条的构图II》《有两条线的构图》绘画的空间关系图2
Fig. 33-4 *Fox Trot A, Composition II with Black Lines, Lozenge Composition with Two Lines*' spatial diagram 2

图33-5 《狐步舞A》《有黑色线条的构图Ⅱ》《有两条线的构图》绘画的空间关系图3
Fig. 33-5 *Fox Trot A, Composition II with Black Lines, Lozenge Composition with Two Lines*' spatial diagram 3

图33-6 《狐步舞A》《有黑色线条的构图Ⅱ》《有两条线的构图》绘画的空间关系图4
Fig. 33-6 *Fox Trot A, Composition II with Black Lines, Lozenge Composition with Two Lines*' spatial diagram 4

图33-7《狐步舞A》《有黑色线条的构图II》《有两条线的构图》绘画的空间关系图5
Fig. 33-7 *Fox Trot A, Composition II with Black Lines, Lozenge Composition with Two Lines*' spatial diagram 5

图33-8《狐步舞A》《有黑色线条的构图Ⅱ》《有两条线的构图》绘画的空间关系图6
Fig. 33-8 *Fox Trot A, Composition II with Black Lines, Lozenge Composition with Two Lines*' spatial diagram 6

思考的延伸
Further thoughts

思考的延伸
Further thoughts

通过上述的35个示例的试做，或许我们已经看出并居然相信了，绘画与建筑之间存在着真正紧密的联系，并可以在一定程度上进行相互转化。

这种绘画与建筑在本质上之所以能够产生重要关联性就在于，绘画和建筑都是人的"意识空间投射的结果"（当然在我个人看来写实性的绘画可能就一定是意识空间投射的结果）。从这个视角和立场（立场很重要）去看每个人所画的线条，其实便不仅仅是二维的图案，而更是一种拥有空间意义的"划痕"。如果肯定一些地讲，这些点与线是具有空间意义的。从构图的层面上讲，平面上的线与线之间讲求的是要关注其所划分的"虚空"的部分。而这又与中国传统中讲求的"留白"相一致，而所谓的"留白"就是空间。从这个意义上，书法中的一笔，与绘画中的一笔，以及建筑师的一条线，其实是一样的，只不过每个人所站的角度和思考范围不一样而已。然而本质上讲，上述的笔与笔之间的距离、长短、大小其实都与创作者头脑中的意识空间相关联，曲线亦如此。

依上述的理解，在绘画中，画面本身所呈现的是画家大脑当中所要表达的一种空间概念的投射。对于

Through the above tests of 35 samples, perhaps we have realised and actually believed there is a real close correlation between painting and architecture, and they can be converted into each other to some extent.

This significant correlation between painting and architecture is, in essence, because they're both "projection of people's consciousness" (to my personal opinion, true-life painting could certainly be projection). From this perspective and standpoint (latter is very important) to see each person's drawing lines, in fact, it is not just a two-dimensional pattern, but a "scratch" with spatial significance. These dots and lines have spatial meanings in certain senses. From the composition perspective, it's the "void" divided by lines that painters focus on. It corresponds with "blank" emphasised in traditional Chinese painting, which is actually the space. In this sense, a stroke in calligraphy or painting, is basically the same as a line of architects' design, only each individual's perspective and consideration are distinctive. Essentially, however, the distance between the strokes, lengths and sizes of strokes, are related to the creators' consciousness, so are curves.

空间概念的本质性理解，绘画与建筑能够得以融合，因为建筑师在进行建筑平面设计的时候正是在对空间进行划分。平面绘画的线条与聚落测绘图纸、音乐中的切分线以及书法等，在对空间的指向性这一点上，都是一种由线和线之间的关系所构成的空间。平面中的线是这样的，立面中的线也是如此。由此看来，绘画与建筑是在线与线、点与点、面与面的层次上获得了统一。

基于这样一种本质上的统一，建筑师便可以从绘画中发现建筑空间。而这种发现的过程，要求建筑师具有柔软的大脑和变幻的视角，以随时调整其对绘画中平面空间的观察的视角：或平面或立面，或大尺度或小尺度……将自己的身体自由地投入画面当中，并于其中变幻，乐莫大焉。这种对空间的思考方式并不局限于绘画中，甚至可以投射于对宇宙中一切事物的观察中。

According to the above-said understanding, what the picture presents in the painting is projection of the painter's spatial concept. The reason painting echoes with architecture lies in that they share the same understanding of spatial concept. When architects design the plan, it's actually space division. Lines in painting, settlement survey drawings, syncopation, and calligraphy, all direct to the space defined by lines. Not only lines in planes do such, but also in facades. Seemingly, architecture and painting achieve uniformity due to layers of lines, dots, and surfaces.

Based on such fundamental unity, architects are able to find architecture space from paintings. And this process of discovery requires architects to have open minds and flexible perspectives, to adjust their angles observing space in paintings at any time. It can be either planes, or facades, large-scale or small-scale… Architects devoting their bodies freely into the painting and changing with them, is great fun. This way of spatial thinking is not limited in just painting, but also almost everything in the universe.

王昀简介

王昀 博士
1985 年毕业于北京建筑工程学院建筑系，获学士学位
1995 年毕业于日本东京大学，获得工学硕士学位
1999 年于日本东京大学获得工学博士学位
2001 年执教于北京大学
2002 年成立方体空间工作室
2013 年于北京建筑大学创立建筑设计艺术研究中心

建筑设计竞赛获奖经历：
1993 年日本《新建筑》第 20 回日新工业建筑设计
　　　竞赛获二等奖
1994 年日本《新建筑》第 4 回 S×L 建筑设计竞赛
　　　获一等奖

主要建筑作品：
善美办公楼门厅增建、60 平方米极小城市、石景山财政局培训中心、庐师山庄、百子湾中学校、百子湾幼儿园、杭州西溪湿地艺术村 H 地块会所等

参加展览：
2004 年 6 月参加 "'状态'中国青年建筑师 8 人展"
2004 年首届中国国际建筑艺术双年展参展
2006 年第二届中国国际建筑艺术双年展参展
2009 年参加在比利时布鲁塞尔举办的 "'心造'—中国当代建筑前沿展"

2010 年参加威尼斯建筑艺术双年展、德国 karlsruhe Chinese Regional Architectural Creation 建筑展
2011 年参加捷克 prague 中国当代建筑展、意大利罗马"向东方——中国建筑景观"展、中国深圳·香港城市建筑双城双年展等
2012 年第 13 届威尼斯国际建筑艺术双年展中国馆参展

WangYunProfile

Dr. Wang Yun
Graduated with a Bachelor's degree from the Department of Architecture at the Beijing Institute of Architectural Engineering in 1985.
Received his Master's degree in Engineering Science from Tokyo University in 1995.
Received a Ph.D. from Tokyo University in 1999.
Taught at Peking University since 2001.
Founded the Aterier Fronti (www.fronti.cn) in 2002.
Prize:
Received the second place prize in the "New Architecture" category at Japan's 20th annual International Architectural Design Competition in 1993.
Awarded the first prize in the "New Architecture" category at Japan's 4th S×L International Architectural Design Competition in 1994.
prominent works:
ShanMei Office Building Foyer, a Small City of 60 Square Meters, the Shijingshan Bureau of Finance Training Center, Lushi Mountain Villa, Baiziwan Middle School, Baiziwan Kindergarten, and block H of the Hangzhou Xixi Wetland Art Village.
Exhibitions:
The 2004 Chinese National Young Architects 8 Man Exhibition, the First China International Architecture Biennale, the Second China International Architecture Biennale in 2006, the "Heart-Made: Cutting-Edge of Chinese Contemporary Architecture" exhibit in Brussels in 2009, the 2010 Architectural Venice Biennale, the Karlsruhe Chinese Regional Architectural Creation exhibition in Germany, the Chinese Contemporary Architecture Exhibition in Prague in 2011, the "Towards the East: Chinese Landscape Architecture" exhibition in Rome, and the Hong Kong-Shenzhen Twin Cities Urban Planning Biennale. The thirteen Venice Architecture-Art Biennale in 2012.

www.fronti.cn